D0153233

From Individual Behaviour to Population Ecology

Oxford Series in Ecology and Evolution
Edited by Robert M. May and Paul H. Harvey

From Individual Behaviour to Population Ecology

William J. Sutherland

School of Biological Sciences
University of East Anglia, Norwich

Oxford   New York   Tokyo
OXFORD UNIVERSITY PRESS
1996

Oxford University Press, Walton Street, Oxford OX2 6DP

Oxford   New York
Athens   Auckland   Bangkok   Bombay
Calcutta   Cape Town   Dar es Salaam   Delhi
Florence   Hong Kong   Istanbul   Karachi
Kuala Lumpur   Madras   Madrid   Melbourne
Mexico City   Nairobi   Paris   Singapore
Taipei   Tokyo   Toronto

and associated companies in
Berlin   Ibadan

Oxford is a trade mark of Oxford University Press

Published in the United States
by Oxford University Press Inc., New York

A catalogue record for this book is available from the British Library

Library of Congress Cataloging in Publication Data
Sutherland, William J.
From individual behaviour to population ecology/William J. Sutherland.
(Oxford series in ecology and evolution)
Includes bibliographical references and indexes.
1. Animal populations.   2. Animal ecology.   3. Animal behaviour.
4. Conservation biology.   I. Title.
QL752.S88   1996   596'.05248—dc20   95-30374

ISBN   0 19 854911 3 (Hbk)
ISBN   0 19 854910 5 (Pbk)

Typeset by AMA Graphics Ltd., Preston, Lancs
Printed and bound in Great Britain by Bookcraft (Bath) Ltd.

Acknowledgements

For comments on all or part of this book I am very grateful to Tim Benton, Peter Berthold, James Deutsch, Paul Dolman, Jenny Gill, Charles Godfray, Paul Harvey, Heribert Hofer, Robert May, Manfred Milinski, Ken Norris, and John Reynolds.

I thank Diane Alden who drew the figures with efficiency and Phil Riley who provided excellent secretarial support. My thanks also go to the staff of Oxford University Press for seeing the book efficiently through production.

Some of the work described here was funded by the Agriculture and Food Research Council, Biotechnological and Biological Sciences Research Council, Joint Nature Conservation Committee, Natural Environmental Research Council, Royal Society for the Protection of Birds, and the University of East Anglia research promotion fund. The wonderful Nuffield Foundation not only gave me a year's employment in 1982 replacing Geoff Parker (who then taught me game theory) but also provided a replacement lecturer last year so giving me time to write this book. The School of Biological Sciences at the University of East Anglia provided a pleasant and intellectually stimulating environment.

Special thanks to Nicola Crockford for reading two drafts and for much else.

Norwich
December 1994   W. J. S.

Acknowledgements

For comments on all or part of this book I am very grateful to Tim Benton, Peter Berthold, James Deutsch, Paul Dolman, Jenny Gill, Charles Godfray, Paul Harvey, Heribert Hofer, Robert May, Manfred Milinski, Ken Norris, and John Reynolds.

I thank Diane Alden who drew the figures with efficiency and Phil Riley who provided excellent secretarial support. My thanks also go to the staff of Oxford University Press for seeing the book efficiently through production.

Some of the work described here was funded by the Agriculture and Food Research Council, Biotechnological and Biological Sciences Research Council, Joint Nature Conservation Committee, Natural Environmental Research Council, Royal Society for the Protection of Birds, and the University of East Anglia research promotion fund. The wonderful Nuffield Foundation not only gave me a year's employment in 1982 replacing Geoff Parker (who then taught me game theory) but also provided a replacement lecturer last year so giving me time to write this book. The School of Biological Sciences at the University of East Anglia provided a pleasant and intellectually stimulating environment.

Special thanks to Nicola Crockford for reading two drafts and for much else.

Norwich
December 1994   W. J. S.

Contents

1

Introduction

1.1 Introduction

Population dynamics and animal behaviour are two subjects that have developed largely in isolation, despite the widespread acceptance that they are related and the frequent acknowledgement of a need to combine them. The major objective of this book is to demonstrate how aspects of the population ecology of vertebrates, such as population size and the response of populations to ecological change, can be related to behaviour. I will concentrate on particular questions which I consider interesting, so this should not be regarded as a comprehensive review of all aspects of these issues.

Studies of the population ecology of vertebrates have tended to concentrate on determining the relevant demographic parameters (Sinclair 1989). Whilst this has been successful, we still do not understand the evolutionary basis to these parameters. A knowledge of the behavioural basis of population ecology has the advantage that, as well as being more intellectually satisfying, it enables extrapolation to novel conditions, for example predicting the responses of populations to habitat loss.

Behavioural ecologists have often examined why animals behave in the manner that they do. The two approaches frequently used for making predictions about behaviour are optimization and game theory. Optimization is used when an animal's behaviour can be considered in isolation (see, for example, MacArthur and Pianka 1966), while game theory is used when the behaviour of one individual is expected to depend upon the behaviour of others (Maynard Smith 1982). Game theory has been successful in the analysis of many aspects of behaviour but has rarely been used to explore the consequences for populations rather than individuals.

Population ecology suffers from having no over all, a priori theory from which explanations and predictions can be devised. Behavioural ecology has such a theory—evolution by means of natural selection—which yields the prediction that individuals will maximize fitness. Thus basing population ecology on behavioural ecology will increase the overall explanatory, and perhaps, predictive power of population ecology.

The links between behaviour and population ecology have attracted interest from various perspectives. Łomnicki (1980, 1988) first emphasized the importance in population ecology of individual differences between animals, although botanists have appreciated the importance for some time (Koyama and Kira 1956). Hassell and May (1985) showed how behaviour and population ecology could be combined in the study of arthropods. These models have been very successful in showing how aspects of behaviour such as interference (see Section 1.6 below) and aggregative behaviour affect population dynamics. However, the framework developed is inappropriate for most vertebrate species for a number of reasons. Much of the elegant theory applies specifically to host–parasitoid systems in which each host encountered by a parasitoid results in one or more eggs being laid, whereas for vertebrates the link between predation and fecundity is more complex (Hassell 1978). Arthropods are usually short lived, and have limited opportunity or intelligence to sample their environment. Vertebrates tend to live longer, have a larger home range and possess greater sampling ability. Finally, many vertebrate species show predictable seasonal migration between wintering sites and breeding sites. Thus, understanding their population dynamics requires an understanding of the pattern of migration. These distinctions are not of course absolute: not all vertebrates show all of these characters while some invertebrates may show some of them.

It is useful to have a conceptual framework for relating field studies and theoretical work, as exists for arthropod predator–prey systems (e.g. Hassell 1978). My hope is to outline such a framework for many aspects of vertebrate ecology, although I hope this will also be of interest to invertebrate ecologists. The approach described here will be useful for species in which the behaviour of an individual is modified by the behaviour of others. I will consider a wide range of taxa but plead guilty to an emphasis on wading birds. My mitigating arguments are that this is the group I know best and they have a relatively simple ecology in which prey density and feeding behaviour can be quantified.

Each chapter of this book will introduce a theoretical framework and illustrate key points with examples. A prime function of theoretical models is often to make clear the relative importance of the components and the priorities for field and laboratory work. They have the advantage of encouraging more rigorous thought; arguments that seem convincing when described verbally or graphically may turn out to be wrong, or true only under unrealistic conditions, when modelled. The models in this book are either general or based on specific systems on which I have worked.

The rest of this chapter will outline the main concepts that will be used throughout the book.

Terminology

In this book I use the term 'patch' to refer to a localized area such as a cockle *Cerastoderma edule* bed or a fruiting tree, and 'site' to refer to a wider area such as an estuary or a wood. In reality there is, of course, a hierarchy of scales ranging from different continents to microhabitats, but incorporating more levels would not affect the arguments presented here. It is conventional to use the term 'habitat' instead of what I call patches or sites, but I will restrict using the term habitat to areas with obvious ecological differences such as different woodland communities. I will use 'consumer' to incorporate insectivores, herbivores, granivores, frugivores, and carnivores and the term 'resources' to include invertebrates, plants, grain, fruit, and other consumers. I will largely avoid the term 'carrying capacity' due to the numerous and conflicting definitions. I will use the term 'sustainable population size' to refer to the maximum number of individuals that can persist within a site.

1.2 Population size

Population size results from a combination of density-dependent and density-independent processes. Figure 1.1 shows an example of density-dependent birth rate and density-independent mortality. The same approach can be used to examine either a combination of density-dependent death rate and a density-independent birth rate or density-dependent death and birth rates. Elucidating the details of behaviour can allow us to explain the shapes of these relationships and hence population size. Thus, one objective of this book is to consider how an understanding of the density-dependent nature of mortality and fecundity can be derived from the behaviour of individuals. A framework for understanding the form of these relationships is provided by the game theory concepts of the ideal free distribution and the ideal despotic distribution (Fretwell and Lucas 1970). Both of these are outlined later in this introduction.

Figure 1.2 illustrates how the behavioural processes such as territoriality, interference and depletion of resources combine to determine population size.

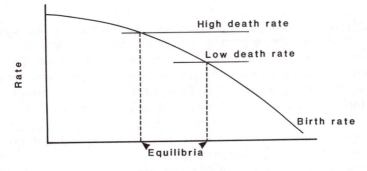

Fig. 1.1 An example of how density-dependent and density-independent processes can be combined to determine equilibrium population size. In this example, birth rate is density dependent. Two density-independent death rates are shown. Thus an increase in the death rate results in a decrease in population size. (Adapted from Williamson 1972.)

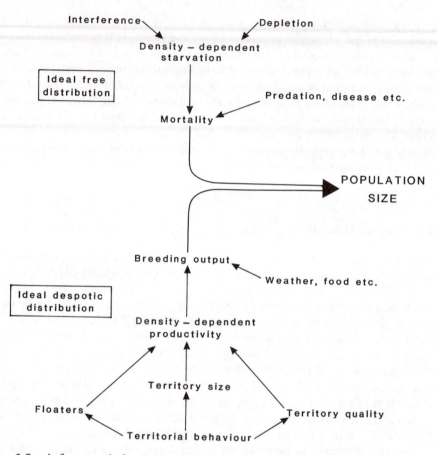

Fig. 1.2 A framework for linking interference and depletion outside the breeding season with territorial behaviour in the breeding season to determine population size.

Large populations result in high depletion and interference (see Section 1.6 below) leading to the possibility of density-dependent starvation. Other processes, such as predation and disease, will also determine mortality rate. In the breeding season territorial behaviour will often influence territory size, the quality of the territory occupied, and the number of individuals failing to breed. All of these are likely to be density dependent. Other density-independent factors, such as the weather, will also influence birth rate. Such density-dependent and density-independent influences on mortality and productivity may then combine to determine population size.

The framework illustrated in Fig. 1.2 can be used to explore the consequences of habitat loss and habitat change for populations. If, say, the area of breeding habitat is reduced, then the density-dependent birth rate may also decline. The consequences of this for the new equilibrium population size can be determined.

1.3 Ideal free distribution

The central idea behind much of this book is that of the ideal free distribution (Fretwell and Lucas 1970) which is described in Fig. 1.3. Ideal refers to the concept that individuals go to the patch where their rewards are highest. Free refers to individuals being able to move where they wish and thus not being constrained by restricted dispersal, territoriality or site-related dominance linked to occupancy. This idea is the same as Parker's (1970, 1974) 'equilibrium position' used for describing the distribution of dungflies *Scatophaga stercoraria* among cow pats. A similar concept was also used by Orians (1969*b*) to account for the distribution of females amongst male territories and by Brown (1969*b*) to account for the choice of woods by great tits *Parus major*.

The obvious consequences of being ideal and free is that the average reward should be the same for individuals in different occupied patches, as shown in Fig. 1.3. If one patch has higher rewards than another, then individuals should keep moving to the better patch until either the rewards balance out or all individuals have left the poorer patch, where their intake would otherwise be lower.

The ideal free distribution is an example of game theory invented before game theory was formally applied to biology. The assumption of game theory is that the gains of adopting one strategy depend upon the strategy adopted by others. This leads to an evolutionarily stable strategy which is defined as a

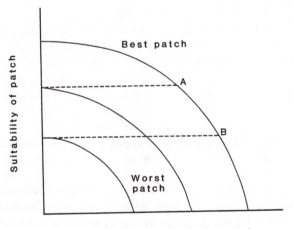

Fig. 1.3 The ideal free distribution. This example shows three patches that, in the absence of competitors, differ in suitability. As the number of consumers occurring within a patch increases the suitability of the patch declines. Individuals will initially settle in the best patch until point A is reached when the suitability in the best patch with many competitors equals that of the intermediate patch with no competitors; both patches will then be used. As more competitors arrive they will settle in both patches until point B when their suitabilities reaches that of the poorest patch, which they will then also start to use. (Adapted from Fretwell and Lucas 1970.)

strategy which, if all the members of a population adopt it, cannot be invaded by a mutant strategy under the influence of natural selection (Maynard Smith 1982). The ideal free distribution is thus an evolutionarily stable strategy as no individual can gain a higher intake by moving. Game theory will be used repeatedly in this book to interpret dispersion, migration patterns, territory size and location, mating systems, and population size.

Dr Fretwell once told me that he never expected the ideal free distribution to be taken particularly seriously. He viewed it as a useful null model to which other factors, such as despotic behaviour, could be added to make it realistic. This is also my view as in almost all cases the simple model proves insufficient and more complex versions are necessary. The ideal free distribution and subsequent modifications have proved influential and, as Milinski and Parker (1991) point out, they have had many applications including the understanding of timing of ontogeny (Werner *et al.* 1983; Werner and Hall 1988), sequential sex changes (Charnov 1992), alternative mating strategies (Parker 1982), Taylor's power law (Gillis *et al.* 1986), metamorphosis (Werner 1986), niche separation (Rosenzweig 1981, 1985), and the spatial pattern of fishing boats (Abrahams and Healey 1990) and whalers (Whitehead and Hope 1991).

1.4 Habitat suitability

An important assumption of the approach taken in this book is that sites and patches differ in their quality. Suitability may ideally be considered as equivalent to fitness. In practice, intake rate and breeding success are often used as substitutes for fitness since they can be measured more readily.

It is reasonable to assume that foraging behaviour is subject to natural selection, although this has rarely been studied. Lemon (1991, 1993) showed that individual zebra finches *Taeniopygia guttato* differ in their ability to discriminate between patches and the better discriminators have a higher fitness as measured in terms of higher fecundity and higher female survival rates. Furthermore, by correlating the patch choice of mothers with their offspring, Lemon (1993) showed that the ability to discriminate between patches was heritable.

Determining suitabilities may be complex; while variation in intake rates will often reflect differences in prey density, other differences may also be important. For herbivores, plant quality is often more important than abundance. The preferences of oystercatchers *Haematopus ostralegus* for different mussel *Mytilus edulis* beds depends upon shell thickness, mussel size, and the muddiness of the substrate (Goss-Custard *et al.* 1992). The risk of being eaten by a predator may also affect the suitabilities of different patches (McNamara and Houston 1987).

The variance in intake rate available from a patch could also play a role in determining choice. Theoretically, whether individuals seek or avoid options that provide a variable reward will depend upon whether or not the mean intake is sufficient. If the average intake is sufficient then individuals will avoid variance in intake as this risks starvation. However, if the average intake is insufficient then variance provides the only possibility of survival (Caraco 1980; Houston and McNamara 1982; McNamara and Houston 1992; Stephens 1981; Stephens and Krebs 1986). The variance experienced by animals in the field may be related to resource size, the extent of prey aggregation, and whether the consumers form flocks (Sutherland and Anderson 1987). Animals with a sufficient intake for survival have been shown in a number of studies to avoid high variance patches (reviewed by Stephens and Krebs 1986). As obtaining sufficient intake must be the usual situation for individuals for most of the time, avoidance of variance in intake is likely to be common. Selecting patches with high variance is likely to be important only when individuals are likely to die from starvation, thus this is probably rarely observed in the field.

In this book I concentrate on biotic interactions, especially the roles of resource depletion and competition between consumers. It is clear that other factors such as temperature, humidity, pH, and salinity are also important in determining where to feed. As just one example, Aldabran giant tortoises *Geochelone gigantea* require shade to avoid overheating and the grazing pattern thus depends upon the distance from shade (Coe *et al.* 1979).

For much of this book I have, for simplicity, assumed a single species of consumer feeding on a single resource. The ideas used in this book can be extended to include more than one species of consumer (Rosenzweig 1986, 1991) or more than one species of resource. In Chapter 10 I will describe models of wildfowl feeding on vegetation, which incorporate more than one species of bird and more than one species of plant.

1.5 Buffer effect

Patches may differ in quality, and the manner in which they are used depends upon the density of competitors. Brown (1969*a*) introduced the term 'buffer effect' to describe his observation that at low population densities individuals tended to occur predominantly in the better patches but at higher densities a larger fraction occurs in poorer patches. Brown's study was of the distribution of tits *Parus* spp. between different areas of woodland. Experimental studies of great tits breeding in Wytham Wood, near Oxford, confirmed this pattern. Those in the wood had a higher reproductive success than those in adjacent hedgerows (Krebs 1971). When Krebs removed territorial birds from the wood their places were then occupied by the hedgerow birds.

The buffer effect has been known for some time (see Rosenzweig 1991) and applies to species in the non-breeding season as well as the breeding season. Figure 1.4 shows a range of examples. The concept of the buffer effect underlies many of the approaches taken in this book.

The buffer effect can act on a range of scales from within patches to across countries. On a local scale Meire and Kuyken (1984) showed that, as the total

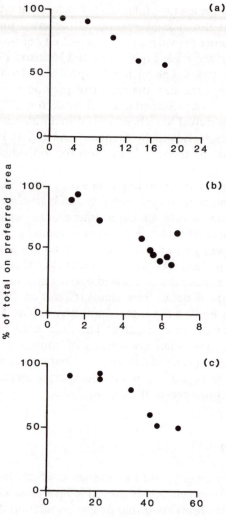

Fig. 1.4 Examples of the buffer effect. As the total population in a site increases the percentage using the preferred patches declines.
(a) Teal *Anas crecca* feeding on seeds (Zwarts 1976); (b) oystercatchers feeding on cockles; (c) knots *Calidris canutus* feeding on the bivalve *Macoma balthica* (Goss-Custard 1977a).

number of oystercatchers increased on a mussel bed, a lower proportion used the parts of the bed favoured initially. At the other extreme, when the British population of grey plovers *Pluvialis squatarola* was low, they concentrated in those estuaries that are warmest in winter (grey plover feed by watching for prey movement and prey are more active at higher temperatures), but as the national population increased the greatest increases in numbers were in the colder estuaries with lower initial grey plover densities (Moser 1988).

1.6 Interference and depletion

A crucial assumption of the ideal free distribution is that as the number of competitors increases within a patch, the suitability declines. In the absence of such negative feedback, and if individual distribution is ideal and free, then all individuals should theoretically occur in the single best patch. Two common forms of negative feedback are interference and depletion.

Interference results in a short-term decline in intake rate due to the presence of others, as a result, for example, of fighting, stealing food, or making prey inaccessible by disturbing it. Depletion is the actual removal of prey. For example, drinking a pub dry would be depletion whilst a crowd around the bar hindering access would be interference.

Interference has also been termed interference competition (Park 1962) and encounter interference (Schoener 1983*a*). Depletion has also been termed exploitation competition (Park 1962; Pianka 1978) and consumptive competition (Schoener 1983*a*).

The manner in which consumers follow the ideal free distribution may differ according to whether interference or depletion is the more important factor. If depletion is important the resources should theoretically be reduced to the same level in all the patches used. If interference is important then the patches with more resources should also have more consumers and the additional interference should result in a constant intake across patches.

The ideal free distribution can be equally applied to other forms of negative feedback (Oksanen *et al.* 1992). For example, high densities of breeding great tits incur high predation (Dunn 1977), and groups of mangabeys *Cercocebus albigena* contaminate the leaves and trunks of their food plants with faeces and then must move on (Freeland 1980). The theoretical expectation is that consumers should distribute themselves such that fitness is maximized, which depends upon a combination of food intake and the risks of predation and disease. Although there is evidence that animals do behave in a manner that balances predation risk and food intake (see Chapter 12) there are not the data to examine whether the behaviour of individuals corresponds with that expected to maximize lifetime fitness.

1.7 Ideal despotic distribution

In species with territoriality or dominance hierarchies the ideal free distribution no longer applies because individuals are not free to move between patches. The first to occupy a patch may gain higher rewards than those arriving later either because the pioneers defend the better areas, or because dominance status

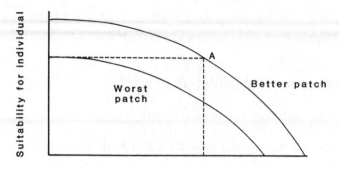

Order of occupancy in patch

Fig. 1.5 The ideal despotic distribution. Due to territorial or dominance behaviour, the suitability of the patch for each individual declines with the order of settling. The presence of others does not decrease the suitability for those that have already settled. Individuals settle in the better patch until point A at which a newcomer would gain equally from settling in the poorer patch. Further individuals will settle in both patches such that suitability for the settling individuals will remain equal. Note that, unlike the ideal free distribution, the average reward differs between patches, although at any time the settling individuals obtain the same reward in both patches. (Adapted from Fretwell 1972.)

may be related to residency. The ideal despotic distribution (Fretwell 1969) can be used to describe the expected distribution between patches (see Fig. 1.5). As shown in the figure, individuals settle in the patch where they gain the highest suitability so that at any time when a number of patches are occupied, the suitability for the next individual is approximately equal across the patches. Unlike the ideal free distribution, however, the mean suitability may vary between patches.

1.8 Allee's principle

Allee's principle is that at very low population density, the rate of survival and reproduction may decline (Allee *et al.* 1949). For example, at extremely low densities, individuals may have difficulties in finding mates. If a lower population density results in a smaller group size then this may affect those predators that hunt more effectively in a group such as predatory fish (Major 1978) and lions *Panthera leo* (Caraco and Wolf 1975). As another example, although the risk of contracting infectious diseases usually increases with group size (Freeland 1976), the eggs of solitary bluegill sunfish *Lepomis macrochirus* were more likely to be infected by fungal pathogens than those of colonially nesting individuals (Côté and Gross 1993).

A major process by which survival or breeding success may increase with density is in avoiding predation. For example, the proportion of lapwing *Vanellus vanellus* clutches lost to avian predators declines with an increase in

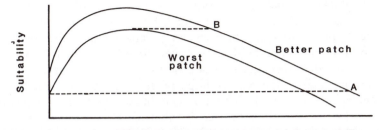

Fig. 1.6 The ideal free distribution with the Allee effect. As in the ideal free distribution (Fig. 1.3), the suitability of the patch declines with the number of competitors but suitability is also low when few competitors are present. Individuals will continue to settle in the better patch until point A where a single newcomer is better alone in the poorer patch. An individual in the poorer patch may increase the suitability of that patch such that individuals then move from the better to the worst patch until they reach point B, where the suitabilities are again equal. (Adapted from Fretwell 1972.)

the number of close neighbours (Berg *et al.* 1992). The obvious response is to breed in colonies and in the lapwing study 92% of the nests were found in such aggregations. It is thought that many species form groups outside the breeding season in order to reduce the risks of predation. Goshawks *Accipiter gentilis* are much more successful when they attack smaller flocks of woodpigeons *Columba palumbus* mainly because the larger flocks can detect the goshawk from a greater distance (Kenward 1978). Similarly, dingoes *Canis familiaris dingo*, a major cause of death of unweaned eastern grey kangaroos *Macropus giganteus*, have a lower probability of suprising larger groups of kangaroos at short range (Jarman and Wright 1993).

Introducing the Allee effect into the ideal free distribution can result in dramatic shifts in theoretical distribution (Fig. 1.6). Fretwell (1986) suggested that this could account for the local fluctuations shown in populations of dickcissels *Spiza americana* in which the survival of nests was greatest at intermediate densities as a result of the combination of greater parasitism by brown-headed cowbirds *Molothrus ater* at lower densities and increased predation at higher nest densities. A similar argument was used by Sibly (1983) to suggest that the optimal flock size may be unstable, although Giraldeau and Gillis (1985) show there are conditions under which group sizes may be stable.

1.9 Sampling

The 'ideal' of the ideal free distribution relates to the assumption that the animals have perfect knowledge of the suitability of each patch, yet this is obviously unrealistic. Individuals clearly have to sample and learn about patches.

The problem of sampling is obviously more complex if the resource population is also changing, for example due to depletion. Bernstein *et al.* (1988) developed a model in which predators depleted the prey population and then moved to another patch. In this model individuals moved if their current intake was less than their memory of past intake. By incorporating psychological learning models, the relative importance of distant or recent intake on the memory of past intake could be modified. With slow depletion the consumers approached the ideal free distribution. However, with rapid depletion the consumers could not learn fast enough to track the environment and no longer conformed to the ideal free distribution. Similarly, if the cost of travelling between patches was large, then consumers became more sedentary, which again led to considerable discrepancies from the ideal free distribution (Bernstein *et al.* 1991*a*).

If searching is costly then the expectation is that individuals will persist in a poorer patch if the additional gain from moving is less than the movement cost, and thus the poorer patches will be overused (Rosenzweig 1981). As an example, in experimental studies of flour beetles *Tribolium confusum* moving between patches differing in resource density, making movement between patches more difficult for the beetles resulted in a higher proportion using the poorer patches (Korona 1990).

The constraints outlined by Bernstein *et al.* (1988, 1991*a*) are likely to have general applications. The ideal free distribution will be less applicable when patch quality fluctuates rapidly, or where it is difficult or costly for the consumer to sample, because for example the distance between patches is great relative to the animal's capacity to move. Animals, however, do make use of long term memory; three-spined sticklebacks *Gasterosteus aculeatus* can remember the quality of patches for at least eight days (Milinski 1994) and individuals use this information in determining where to feed within the ideal free distribution.

1.10 Scale

Behaviour and population ecology at one scale may differ from those at another (Wiens 1989). For example, individuals may learn which patch to feed in, whereas the choice of wintering area may be under genetic control (Berthold *et al.* 1992). It is probably generally true that individuals are able to sample, to some degree, patches within a site but sampling of different sites will be relatively more restricted.

The size of an area may influence both mortality rate and breeding success. Fitzgerald *et al.* (1992) studied three-spined sticklebacks that had settled at similar densities in different sized ponds. More fish were present in large ponds and this resulted in an increase in egg cannibalism and a decline in reproductive success. Similarly, studies of shelduck *Tadorna tadorna* showed that in larger sites there were more pairs and more territorial interactions. Breeding success was lower in these larger sites as chicks regularly died as a result of territorial interactions (Pienkowski and Evans 1982). It thus does not necessarily follow that processes observed at one scale apply when considering other scales.

1.11 Structure of the book

Each chapter presents a combination of theory and empirical examples drawn from the literature. The theory outlined in detail is either my own or has been carried out in collaboration with colleagues. My objective has been to extend the theory with examples rather than provide a comprehensive literature review.

Chapters 2 and 3 show how game theory models can be used to describe the distribution of individuals in relation to resources in the presence of interference and depletion respectively. Chapter 4 shows that considering the variation in prey availability is important in understanding many ecological issues such as consumer aggregation, functional responses, and prey mortality. Chapter 5 uses the theory of the ideal despotic distribution to describe territorial behaviour. This theory examines where individuals should settle taking into consideration the distribution of other territorial individuals and the distribution of resources. This can be used to consider the nature of density-dependent fecundity, delayed breeding, and co-operative breeding.

Chapter 6 shows that these theories, developed to account for the distribution of consumers in relation to their food, can also help in understanding the distribution of males among breeding sites—in this case leks. This can provide an explanation for female selection for larger leks.

Chapter 7 examines population regulation. The models in previous chapters describe how intake depends upon the distribution of resources, the number of consumers, the level of interference, and the depletion rate. By incorporating a threshold intake necessary for survival it is possible to predict how many individuals will starve and thus determine the nature of density-dependent mortality. This can be used to examine the equilibrium population size for a single wintering site and a single breeding site. Chapter 8 extends the discussion to consider a migrant facing choices between a range of sites and predict evolutionarily stable migration routes and resulting equilibrium population sizes. Chapter 9 considers the pattern of prey mortality expected from the consumer distribution.

The aim of Chapter 10 is to show how these general models can be used to describe specific situations. The models are then used to consider a range of applied questions such as predicting the consequences of habitat changes on species of conservation interest. Habitat loss is a worldwide problem and Chapter 11 shows how the framework presented throughout the book can be used to predict the consequences of habitat loss for both sedentary and migratory populations and reviews some examples. Conservationists are often concerned with the impact of disturbance on animal distribution and Chapter 12 incorporates disturbance into the framework. Disturbance can be considered as equivalent to the risk of predation. Hence this chapter considers how consumers choose feeding sites when faced with a trade-off between food supply and predators. It is then possible to predict the consequences of disturbance for population size. Chapter 13 describes the main techniques used in creating the models outlined in this book.

1.12 Summary

The objective of this book is to provide a framework for studying the consequences of individual behaviour for vertebrate populations. This chapter outlines the basic ideas. The starting point is the concept of the ideal free distribution, whose assumption is that individuals should choose to occupy the patch where suitability is highest. As a consequence of the negative feedback caused by interference and depletion, suitability should be constant in all patches used by the consumer. The ideal free distribution can be modified by assuming that individuals can monopolize resources (the ideal despotic distribution) or by including the 'Allee effect' in which individuals gain from the presence of others, for example by a reduced predation risk when feeding in groups. Also considered are the difficulties individuals face in sampling patches for suitability and the fact that the effect of the processes discussed here may vary with the scale at which they are examined.

2

Interference

2.1 Introduction

Interference is the decline in resource use resulting from the behaviour of other individuals. This does not include depletion, in which the resource is removed. Entomologists have long known that parasitoids at high densities are less efficient at finding hosts (Hassell 1978) but there has been much less work studying the role of interference in vertebrates. In this chapter I will consider the process of interference and its consequences in determining the distribution of individuals.

There are a number of ways in which interference may occur amongst vertebrates (Goss-Custard 1980). Fighting may increase with consumer density. For example, knots fight more when close together (Goss-Custard 1977a). Some species regularly have their prey stolen by another species, i.e. klepto-parasitism (see Section 2.7) and the incidence of this may increase (or decrease) with density. Prey disturbance is another source of interference, i.e. the presence of competition may cause prey to hide.

2.2 Quantifying interference

The importance of interference can be shown by plotting mean feeding rate against consumer density. Figure 2.1 shows how the average number of mussels taken by oystercatchers declines with the number of oystercatchers nearby (Sutherland and Koene 1982). This is reversible—the intake rate declines at high bird densities but returns to previous levels if the bird density is low again. Thus this fall in intake rate is clearly not due to depletion. A major reason for this interference is that oystercatchers steal mussels from each other and the frequency of this increases with bird density (Goss-Custard *et al.* 1984). As explained below, this feeding rate is expressed in terms of search time—the total time spent foraging minus the total time spent handling prey.

Following Hassell and Varley (1969) interference is usually expressed by

$$a'_i = Q\, P_i^{-m} \tag{2.1}$$

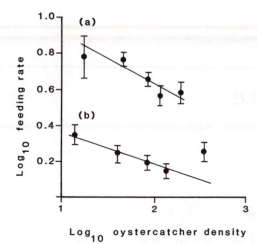

Fig. 2.1 The interference experienced by oystercatchers in two studies expressed as the decline in feeding rate with an increase in the density of competitors. (a) Schiermonikoog, Netherlands (Zwarts and Drent 1981), (b) Texel, Netherlands (Koene 1978). (From Sutherland and Koene 1982.)

where a'_i is the searching efficiency of a consumer in patch i, P_i is the consumer density in patch i, Q is the Quest constant (the value of a' achieved by an individual feeding alone) and m is the interference constant. Thus a high value of m indicates that searching efficiency declines markedly with consumer density whilst a low value indicates that interference is less important.

Values of m can be calculated from field data as the slope of the logarithm of the intake rate expressed in terms of time spent searching (i.e. removing the total time spent handling prey) against the logarithm of the consumer density as shown in Fig. 2.1. If total handling time is not removed from the measure of intake rate then this will underestimate the value of m, however, if handling time is low or if items are caught infrequently then these two measures would be similar.

Although the interference constant m has been estimated for many invertebrates, there are very few values of m for vertebrates. For oystercatchers feeding on mussels (Fig. 2.1) the estimated values of m were 0.10 ± 0.04 and 0.35 ± 0.11 (Sutherland and Koene 1982). The values of m for arthropods range between 0 and 1.13 but are considered overestimates (Hassell 1978), although there may also be biases resulting in underestimation (Arditi and Akçakaya 1990). A further complexity is that this approach assumes that the interference coefficient is constant across the range of consumer densities although its value is likely to increase with consumer density (Arditi *et al.* 1992; Free *et al.* 1977).

In many species interference may be trivial. This is particularly likely for species feeding on an immobile resource, which obviously cannot be disturbed. Thus Goss-Custard (1970) showed that redshank *Tringa totanus* experience interference when feeding upon the amphipod *Corophium volutator*, which scuttles down its burrow when a redshank walks over, but not when feeding on the snail *Hydrobia ulvae*. Interference may be lower for a low-value resource that is not worth fighting over or a resource that can be swallowed rapidly before an attack is possible. For example, when oystercatchers fed on cockles, which are much smaller and quicker to open than mussels, there was no evidence of any interference (Sutherland and Koene 1982).

2.3 Interference and the ideal free distribution

Two conflicting processes affect intake rate. Firstly, a resource will usually be easier to find when it is at a high density and thus consumers will favour patches of higher resource density. Secondly, interference increases with the density of consumers and thus also increases with resource density. The ideal free distribution describes the point at which these two conflicting processes balance out so that the intake is equal over all patches. This can be solved numerically by inserting the description of interference given in eqn 2.1 into Holling's (1959) disc equation (see Section 3.2), which describes how intake increases with resource density, to give an intake rate in relation to both resource density and interference. The resulting equation is horrible. However assuming the ideal free distribution is true, and thus intake rate is constant in each patch, the equations simplify to give the following relationship between the number of consumers P_i and the number of resource items N_i in a patch

$$P_i = c \, N_i^{\,1/m} \tag{2.2}$$

where c is a normalising constant (Sutherland 1983). If $m = 0$, then all individuals feed in the patch with the highest resource density.

Thus, this equation describes the distribution of consumers at which the differences in resource density are balanced by interference so that the intake is equal in all the occupied patches. In addition, there may be patches of lower resource density in which the intake would be lower and thus should be ignored. As the consumer density increases, higher interference will be experienced in the richest patches resulting in the use of a wider range of patches.

This approach can then be used to describe the numerical response, that is the relationship between the densities of consumers and their resource (Solomon 1949). Figure 2.2 shows the aggregative response expected from the ideal free distribution derived from eqn 2.2 with four different values of interference and in the absence of depletion. The result is intuitively obvious: with weak interference most consumers collect in the patches of highest prey density but with stronger interference they are spread more evenly between patches. An example of this process is the redshank mentioned in the previous section. When they fed on the gastropod *Hydrobia ulvae* and interference was negligible, the birds fed in dense flocks in the richest patches. However, redshank feeding on the amphipod *Corophium volutator* experienced interference and thus the birds were dispersed (Goss-Custard 1970).

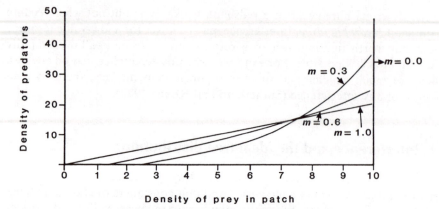

Fig. 2.2 The expected aggregative responses (from eqn 2.2), with different levels of interference, from the ideal free distribution.

This model (eqn 2.2) uses the simplest form of the interference equation (Hassell and Varley 1969). The main limitation of this model is that it assumes that the interference constant does not change with the density of consumers. Although we know little about interference in vertebrates, there is empirical evidence for invertebrates that the interference constant increases with density. A number of models of parasitoids have been suggested that incorporate such an increase (for example Beddington 1975). If we assume that the interference constant increases with the number of consumers, the consequence will be that as the population increases, a greater proportion will occur in the patches with lower resources. Thus, this is a possible explanation of the buffer effect, explained in Chapter 1, in which at low densities most individuals use the best patches but as the density increases a higher proportion use the poorer patches. It is clear that we need more empirical field studies to improve our understanding of interference amongst vertebrates.

Table 2.1 reviews all the studies I know where the observations of foraging animals are tested in relation to the ideal free distribution. Those studies in which resources are continually and rapidly replenished, as in the experiments in which food is added to two patches at different rates, are considered separately in Section 3.8. The expectation of the ideal free distribution is that intake will be constant in all patches, yet this is clearly not the norm; in almost all cases the individuals in the most preferred patch had the higher intake. Although the ideal free distribution may provide a useful theoretical basis it is clear that it is insufficient to account for the patterns observed. One clue as to the cause of the discrepancy may be that most of these studies recorded that individuals differed in competitive ability.

Table 2.1. Summary of field studies on vertebrates relating to the ideal free distribution of feeding individuals, comparing the intake on different patches; whether the observations agree with the prediction of the ideal free distribution for identical competitors that the intake should be equal in all sites is also shown. (From Parker and Sutherland 1986.)

Species	Result	Agreement with ideal free model for identical competitors	Reference
Shelduck *Tadorna tadorna*	Less time spent feeding in dense prey patches	Possibly contradicts model	Buxton (1981)
Oystercatcher *Haematopus ostralegus*	Average intake differs consistently between mussel beds	Contradicts model	Goss-Custard *et al.*(1984)
Herring gull *Larus argentatus*	Average intake about five times greater in better area of rubbish tip	Contradicts model	Monaghan (1980)
Herring gull *Larus argentatus*	Average intake consistently higher on open tip than elsewhere	Contradicts model	Sibly and McCleery (1983)
Oystercatcher *Haematopus ostralegus*	Average intake differs between parts of cockle bed	Contradicts model	Sutherland (1982*c*)
Teal *Anas crecca*	Feeding duration independent of prey density	Possibly supports model	Tamisier (1974)
Shelduck *Tadorna tadorna*	Time spent feeding similar between areas with different prey densities	Weak support for model	Thompson (1981)
Lapwing *Vanellus vanellus* and Golden Plover *Pluvialis apricaria*	Rate of intake greater in fields where prey is most abundant	Contradicts model	Thompson (1984)
Oystercatcher *Haematopus ostralegus*	Intake constant in years of different mussel availability	Weak support for model	Zwarts and Drent (1981)

2.4 Differences in competitive ability

The ideal free distribution assumes all individuals are equal, yet as Darwin (1859) stated 'no one supposes that all individuals of the same species are cast in the very same mould'. There is evidence for differences in competitive ability caused by factors such as size, age, or aggressiveness between fish (e.g. Magurran 1986), amphibians (e.g. Wilbur 1984), reptiles (e.g. Stamps 1991), birds (e.g. Whitfield 1990), and mammals (e.g. Bell 1986). Competitive ability will often be closely related to fighting ability but this need not be the only basis.

Under competition there may be increased differences in the amount of resource individuals acquire. The everglades pygmy sunfish *Elassoma evergladei* shows considerably greater individual differences in growth rate whilst under competition than when in isolation (Rubinstein 1981). Similar results have been obtained for tadpoles (Wilbur and Collins 1973).

Ens and Goss-Custard (1984) quantified the extent to which individual oystercatchers experience interference. Some birds regularly attacked others, and were rarely attacked themselves. The aggressive birds gained about 20% of their intake by stealing mussels whilst subdominants lost a considerable amount of their prey to other individuals. The intake of dominants was unaffected by high densities and some individuals even increased their intake as there were more birds from which to steal mussels. By contrast, the intake of subdominants declined with density. This could partly be attributed to higher losses due to stealing but much of the decline is due to them simply finding fewer mussels. The reasons for this are unknown but they may have been distracted by the presence of many potential attackers or may have fed in the poorer part of the mussel beds.

The effects of individual differences in competitive ability on intake can be incorporated in the relationship between intake and competition density as either differences in the slope or as differences in the intercept (Fig. 2.3). Differences in slope reflect individuals differing in their susceptibility to interference. If individuals differ in their ability to find resources then this will be reflected in their differences in intercept.

This individual variation is incorporated into the model by assuming that the population consists of a range of competitive classes each with a different competitive ability. These individual differences in competitive ability can be incorporated into the model of interference by assuming that the searching efficiency for a member of competitive class s in patch i is

$$a'_{(s,i)} = Q\,P_i^{-mR\,(s,i)} \tag{2.3}$$

where R is the relative competitive ability expressed as the mean competitive ability of all other individuals in that patch divided by the individual's competitive ability (Parker and Sutherland 1986; Sutherland and Parker 1985). Thus, a poor competitor will experience considerable interference when

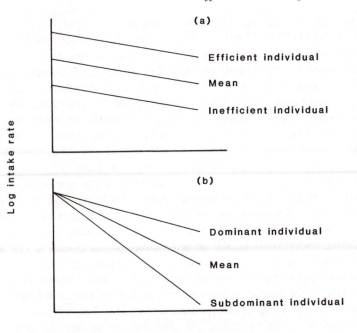

Fig. 2.3 The manner by which individuals may differ in the interference they experience: (a) individuals with different searching efficiencies, such as would occur if individuals differed in their ability to find resources; (b) individuals experiencing different interference, such as would occur if individuals differed in their dominance.

surrounded by good competitors yet only average interference when surrounded by other poor competitors.

The aim is then to seek the strategy by which individuals of each competitive class obtain the highest intake. The distribution between patches may be predicted by taking into account the behaviour of others of the same and different competitive classes, as in Parker's (1982) phenotype-limited evolutionarily stable strategy. This is the distribution at which no individual of any competitive class gains by moving. As a result all members of the same class should have the same intake—as otherwise they would move to take advantage of a higher intake elsewhere. There are, of course, differences in intake between competitive classes.

The rules for the occupancy of different patches can be solved numerically, but the distribution is most easily determined by simulating a population possessing a range of values of competitive ability. An easy way of determining the evolutionarily stable strategy is to calculate the intake rate of each competitive class in each patch, multiply the number of each class in each patch by their relative intake rate, and then repeat these steps until the distribution is stable and the intake rates for each class are equal in all occupied patches.

By this process, if for one class the intake is higher in one patch than another, then that class will shift until either all individuals have abandoned the patch providing the lower intake or the intakes become equal due to the change in interference caused by individuals moving. As a result, each competitive class moves until its intake is the same in all the patches it inhabits. As all classes are shifting simultaneously the final solution gives the evolutionarily stable strategy for each competitive class taking into consideration the behaviour of all other classes. The same final result is obtained regardless of the initial distribution.

Figure 2.4 shows an example of the resulting distribution. The theoretical expectation is that the good competitors accumulate in the richest patches and gain the highest intake. Weak competitors tolerate a low intake in poor sites because if they moved to a richer site they would suffer greater interference and have an even lower intake. This may help explain the widespread observation that individuals persist in poor patches and obtain a low intake but do not move elsewhere (see Table 2.1). Incorporating such individual differences in competitive ability can also help explain the common observation that individuals in the better patches have a higher intake, more fights, and are more likely to be of higher competitive ability (see Table 2.1).

As one example of this process, the distribution of herring gulls *Larus argentatus* was studied in two patches on a rubbish tip: a main area in which the rubbish was dumped directly and a secondary area surrounding this (Monaghan 1980). The density of gulls was higher in the main area and their

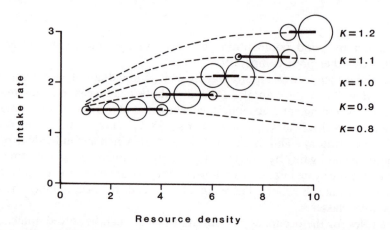

Fig. 2.4 The distribution between patches for individuals differing in competitive ability. The population consists of 500 individuals divided equally into five different phenotypes that differ in competitive ability (K). The resource density increases linearly across 10 patches. The dashed lines show the intake each phenotype would gain in each patch at equilibrium. Each individual feeds where its intake is highest. The number of each phenotype in each patch is indicated by the radius of the circle. Average interference, m (see eqn 2.3) = 0.3.

average intake was about five times higher, but there were also three times as many aggressive encounters. The secondary area was used more by immatures when the total number of gulls was high.

Similarly, a number of studies have shown how adults and juveniles differ in distribution in the manner expected by this model. The most preferred mussel beds have a lower proportion of juvenile oystercatchers (Goss-Custard *et al.* 1982*b*) and areas with the most dunlin *Calidris alpina* have the lowest percentage of juveniles (van der Have *et al.* 1984). Although there are other possible explanations for these observations, for example juveniles may sample more widely and be less experienced or able at choosing patches, in Goss-Custard's study it was clear that juvenile oystercatchers were leaving beds because of the interference from adults. The juveniles occupied the preferred beds in the summer when the adults were away on the breeding grounds, but abandoned these beds when the adults returned.

The same process may influence the distribution of the sexes. Downy woodpeckers *Picoides pubescens* show sex differences in foraging behaviour with the females feeding lower in the canopy and foraging more on tree trunks and branches. When the males were experimentally removed the females then adopted the niche previously occupied by the males (Peters and Grubb 1983).

The unequal distribution of individuals differing in competitive ability may also occur over a greater scale. Swennen (1984) compared the composition of a number of roosts of oystercatchers on the Dutch Waddensee. These roosts differed according to the extent of adjacent intertidal mud-flats. In the roosts with the smallest mud-flats there were fewer birds and a higher proportion of juveniles, immatures, and males (which are smaller than females). There was other evidence that these roosts contained birds of poorer quality—birds of each age class were, on average, below their expected weight and there were also more birds with deformed bills. During a period of severe weather a greater proportion of the birds from these roosts died. This study was carried out on two islands on the Waddensee and the same pattern was shown in each.

As shown in Fig. 2.4, the theoretical expectation from models incorporating differences in competition ability (Parker and Sutherland 1986; Sutherland and Parker 1992) is that distributions are truncated—competitive classes do not overlap by more than one patch. This rule is proved analytically by Parker and Sutherland (1986).

It would, of course be naive to expect such a narrow exploitation of patches by individuals in the field. For example, individuals must sample patches and this will confound such a pattern. Even if consumers were to fit the expectations perfectly, there are two processes which will blur such a truncated pattern and make it hard to detect. Firstly, for apparently recognizable classes, there may well be some overlap in competitive ability. For example, some juveniles may be better competitors than some adults. Secondly, the researcher and the study animal may differ in the way they classify patches. Thus, poor competitors may occupy the poorer areas within what is otherwise a good-quality patch, and the observer might interpret this as different competitors coexisting.

Individual differences in intercept, caused by differences in foraging ability, as shown in Fig. 2.3, are not expected to influence where individuals feed (Sutherland and Parker 1992). It is, however, likely that the poor competitors, which experience more interference, will also be the poorer foragers. This may well increase the differences in mean intake between patches and make it more likely that the individuals in the poorer patches will starve.

2.5 Functional responses and aggregative responses

It has long been recognized that there are two crucial components to the population dynamics of consumers and their resources. The aggregative response is the relationship between the numbers of consumers and resource density in different patches. The functional response is the relationship between the number of prey eaten and resource density (Solomon 1949).

In practice, functional responses are measured in two different ways. One is to place captive animals in different resource densities and determine the amount of food eaten. This measures the intake rate including the time individuals are not feeding, for example because they are satiated. The alternative approach, used largely in field studies of birds, is to measure the intake rate of birds foraging in patches of known resource density. In such studies handling time is usually measured directly as the time between encountering one item and resuming searching after swallowing it. This then excludes birds that are resting or satiated. The interpretation of handling time depends on the method. If the daily intake rate is measured, then the handling time includes time spent not feeding as a consequence of satiation.

In this section I show how the ideal free distribution provides insights into the shape of both the functional and aggregative responses. The approach is to consider a range of patches and use eqn 2.3, relating intake to interference and competitive ability, to determine the evolutionarily stable strategy of who feeds where. The number of individuals in each patch and their mean intake is then calculated (Sutherland 1992).

2.5.1 *Aggregative responses*

A series of aggregative responses for different levels of interference are shown in Fig. 2.5. If all individuals are equal then the results from these simulations are identical to those derived from eqn 2.2 and shown in Fig. 2.2.

Variation in competitive ability can then be incorporated by considering a range of competitive classes. Figure 2.5 shows the aggregative response for differing extents of variation in competitive abilities. With greater variation in competitive ability, poor competitors suffer stronger competition in the richest patches and thus exploit the other patches. With high levels of variation the richest patch may even have fewer individuals than an intermediate patch; most individuals avoid the richest patches as they would experience very high

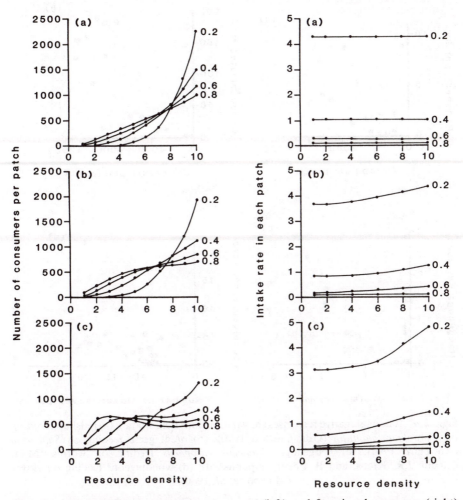

Fig. 2.5 The predicted aggregative responses (left) and functional responses (right) with different degrees of variation in competitive intake and interference. Each graph shows four levels of interference, *m*. In (a) all individuals are equal, in (b) competitive ability varies between 0.9 and 1.1, and in (c) competitive ability varies between 0.8 and 1.2.

interference from the high competitive ability individuals present there. Thus the highest densities need not be in the richest patches, as has been shown in field studies (Van Horne 1981).

A range of actual aggregative responses are shown in Fig. 2.6. In these studies, individuals tend to occupy patches which vary considerably in resource density. If values of interference are low, then all individuals should crowd into the few richest patches. We have little idea of typical values of interference, *m*. The best estimates, of 0.1 and 0.35, are for oystercatchers (see Fig. 2.1) but

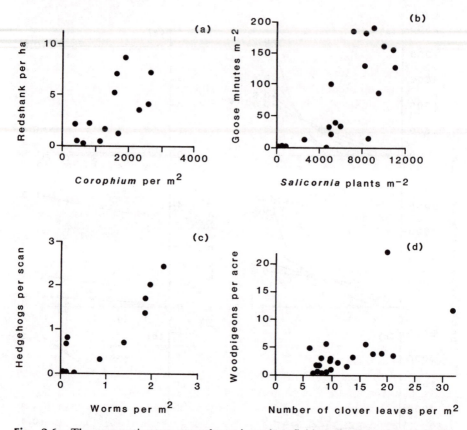

Fig. 2.6 The aggregative responses shown in various field studies: (a) redshank feeding on *Corophium volutator* (Goss-Custard 1970); (b) brent geese feeding on *Salicornia* (Rowcliffe 1994); (c) hedgehogs *Erinaceus europaeus* feeding on worms (M.H. Cassini, J.R. Krebs, and B. Föger, unpublished); (d) woodpigeons feeding on clover (Murton *et al.* 1966).

these may be higher than average as this species is particularly prone to stealing food from each other. Further estimates are clearly needed. If interference is relatively low and insufficient to explain why they use poor patches then it may well be that variation in competitive ability is one explanation of this.

2.5.2 *Functional responses*

By calculating the mean intake of all individuals in each patch it is possible to determine the functional response. These are shown in the right hand graphs in Fig. 2.5. The top graph shows the case with the ideal free distribution in which all individuals are equal and the intake is thus the same in all occupied patches. Incorporating individual differences shows that the better competitors have a higher intake rate and occupy the sites of higher resource density. As a result mean intake rate increases with resource density. The slope of the functional response increases with increasing variation in competitive ability.

Many field studies have shown that the mean intake rate is highest in the patches with the most resource. Figure 2.7 shows the actual functional responses for a number of field studies of birds. The obvious question is why do more individuals not go to the patches with the highest resource density. This approach shows how individuals with a low intake in the poor patches would not necessarily gain by moving to richer patches.

The traditional explanation for the shape of functional responses is that the upper limit is determined by the handling time of food items (Holling 1959). If this is true then the maximum intake rate should be the reciprocal of the handling time. However the value of the handling time necessary to produce the intercept of such a relationship is usually much lower than the values of handling time actually observed. For example, the observed handling time for skylarks *Alauda arvensis* is 1–2 seconds while it would have to be about 12 seconds to produce the observed intake rate (Green 1989). Oystercatchers studied by Goss-Custard (1977a) would have to have a handling time of about 55 seconds although the observed time is 19–29 seconds (Sutherland 1982a). It is clear that although handling time plays a role it is insufficient to account for the level at which the asymptote occurs. Figure 2.5 shows one explanation; the intake rate may stay reasonably constant over a range of prey densities because of the balance between interference and resource density.

My aim in this chapter has been to point out the consequences of interference and variation in competitive ability for functional and aggregative responses. Many other factors obviously play a role, for example, the functional response of many herbivores may be limited by their gut capacity, and secondary chemicals may influence digestion in complex ways (Crawley and Krebs 1992). Changes in diet selection can also influence the shape of the functional response (Krebs *et al.* 1983). In Chapters 3 and 4, I will show how depletion and variation in resource availability can also be important components. However, from the models in this chapter it appears that in order to understand the shape of the aggregative and functional responses in the field it may be also necessary to consider interference and the degree of inequality between individuals.

2.6 Dominance hierarchies

In some species there may be a clear linear dominance hierarchy. This is particularly likely for organisms with small home ranges in which they regularly encounter, and thus can recognize, the neighbouring individuals, or at least their status.

Dunnocks *Prunella modularis* usually feed alone during the winter but may aggregate temporarily in patches, especially during periods of snow cover when only a few feeding patches are exposed. Davies (1992) created patches of food and showed that there was a clear linear dominance hierarchy at these patches

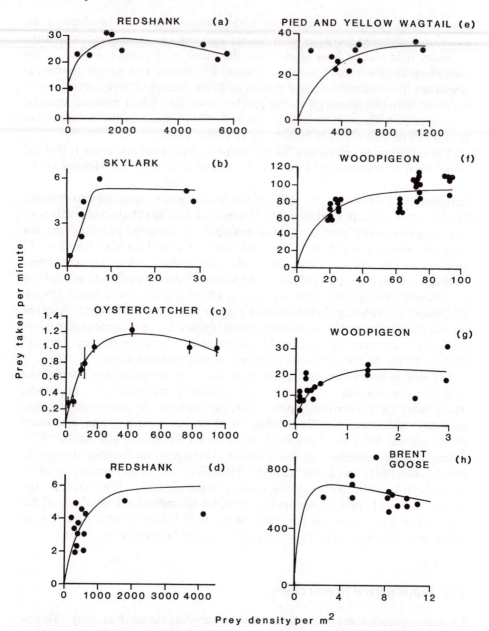

Fig. 2.7 The functional responses shown in various field studies. All prey are expressed as number per m⁻²; (a) redshanks feeding on *Corophium volutator* (Goss-Custard 1977*a*); (b) skylarks *Alauda arvensis* feeding on weed seeds (Green *et al.* 1978); (c) oystercatchers feeding on cockles (Goss-Custard 1977*c*); (d) redshanks feeding on worms (data of Goss-Custard (1977*b*); (e) wagtails *Motacilla* spp. feeding on flies (trap totals) (data of Davies 1977); (f) woodpigeons feeding on clover leaves (Murton *et al.* 1966); (g) wood pigeons feeding on cereal grains (Murton *et al.* 1964); (h) brent geese feeding on *Salicornia* (seeds digested per minute) (Rowcliffe 1994).

with males dominant to females. Dominant birds spent more time feeding and on the two occasions in which dominant males were killed by cats, the females then spent more time at the feeder. In addition to being subdominant and thus being excluded, the subdominant dunnocks were also the last to leave when disturbed by either a predator or by Davies, implying that they were the most vulnerable to predation. Their exclusion from patches and their increased predation risk are thought to be responsible for the increased mortality of females in cold weather. After severe winters the population can be significantly biased towards males (Davies 1992).

Dominance hierarchies can be used for determining how many individuals an area can sustain. Spotted hyaenas *Crocuta crocuta* show a clear dominance hierarchy on the kill (Frank 1986); the dominant individuals have greater access and spend more time feeding than subdominants. Heribert Hofer and I used Frank's relationship between rank and intake in an attempt to determine the number of hyaenas that could be sustained within a territory. This model included an estimate, based on field observations in the Serengeti, of the number of prey that the clan could catch, both in the dry season when most of the prey have migrated elsewhere, and in the wet season when prey are abundant.

When the dominant hyaenas were satiated and stopped feeding the subdominants all rose in rank on any kills and thus were more likely to feed. The prediction of the model was that during the wet season all individuals could feed on the territory but that during the dry season the low-ranking individuals could not obtain sufficient intake to stay alive. In practice, the low-ranking individuals that obtained insufficient resources left the territory on foraging excursions to hunt migrating game (Hofer and East 1993). Thus this simple model showed how rank and food supply could interact to help interpret the number of animals that could be sustained, how this varied with season, and which individuals would gain insufficient food.

2.7 Kleptoparasitism

Kleptoparasitism, the stealing of resource, is widespread (Brockmann and Barnard 1979). Food stealing has been observed in a wide range of species including red-eared sliders *Pseudemys scripta* (Hayes 1987), crested caracara *Polyborus plancus* (Rodríguez-Estrella and Rivera-Rodríguez 1992) and spotted hyaenas (Kruuk 1972). Interference may act through interspecific kleptoparasitism but here I will consider only intraspecific kleptoparasitism. There are a number of models of kleptoparasitism, each of which differs in the precise assumptions made and its conclusions. It is thus not yet clear what the expected consequences of kleptoparasitism are.

In some species, individuals specialize in being either producers or scroungers (Barnard 1984). In other species, individuals cannot be clearly classified as either parasites or victims. Amongst oystercatchers all individuals find resources for themselves, but the more dominant an individual is, the more likely it is to steal resource from subdominants (Goss-Custard *et al.* 1982*a*). Individual spice finches *Lonchura punctulata* are able to be either producers or scroungers and adjust their use of each strategy in accordance with the relative benefits of each in a given situation (Giraldeau *et al.* 1994).

The models of Beddington (1975) and DeAngelis *et al.* (1975) describe the interference occurring due to kleptoparasitism, making the assumption that whenever consumers meet they fight. This does not, however, seem the norm in the field, and at least in oystercatchers, leads to a predicted value of the interference coefficient that is about a million times smaller than those actually observed (Sutherland and Koene 1982).

Holmgren (1995) provides a model of the distribution of kleptoparasites that seems to relate well to the real world. Holmgren's model works by assuming that at any given time each foraging individual can be classified as either searching, handling prey, or fighting. The transitions between these three behaviours can then be described by a set of differential equations. The population comprises individuals differing in rank. If a dominant encounters a subdominant handling a resource item it will fight and steal the resource. After fighting for a certain time over a prey item, the winner continues to handle the prey and the loser starts searching. In this model there is a choice of two patches which differ in prey density. Individuals feed in the patch in which their intake is highest.

Holmgren's model provides sensible results in which the richest patch has a greater number of individuals and a higher proportion of dominants. It predicts that both dominants and subordinates occur in both patches. The top individuals occur in the richer patch but after that individuals of decreasing rank tend to alternate between patches. As a result, mean intake will differ between patches as shown in field studies (Table 2.1).

A clever aspect of this model is that it does not assume a certain level of interference but instead interference arises from the decline in intake due to fighting. Indeed it would be possible to determine the values of the interference constant for different conditions such as the time spent in each fight.

Hassell and Varley's (1969) interference model has been subject to criticism as it assumes that the intensity of interference is independent of competitor density (Arditi *et al.* 1992; Free *et al.* 1977). An advantage of Holmgren's model is that the extent of fighting inevitably increases with competitor density.

The various models of the ideal free distribution incorporating kleptoparasitism give a range of predicted distributions. In Parker and Sutherland's (1986) model, individuals were either kleptoparasites or were victims. Such a model does not result in a stable equilibrium—as the kleptoparasites move to the patches with the most victims, the victims move away from them. Korona (1989) provides a model of the ideal free distribution with unequal competitors based on the assumption that any two individuals will compete between them over food items. The probability of getting the resource is proportional to the competitive abilities of the two rivals. Like Holmgren's model, this predicted that both dominant and subdominant individuals occur in both patches. More field studies are required to determine the precise manner by which such competition usually operates before it can be determined which model is the most sensible.

2.8 Summary

Interference is the decline in intake resulting from the behaviour of other individuals. Simple models of the ideal free distribution incorporating interference can be created, but these fail to describe the field observations of consumers. The models can be made more realistic by incorporating individual differences in competitive ability. Such models seem more successful at accounting for the distribution of individuals between patches. In the richer patches there are more competitors and more individuals of higher competitive ability, but the average intake still exceeds that of the poorer patches. This pattern is shown in many field studies. These models can then be used to predict functional and numerical responses. Some species may have clear dominance hierarchies with the low-ranking individuals occupying the poorer patches and being more prone to starvation. With an understanding of the relationship between dominance and intake it is possible to explain the number of individuals a site can sustain.

3

Depletion

3.1. Introduction

Depletion, the removal of resources by consumers, is undoubtedly an important process in determining the distribution and intake of individuals. Depletion will always occur where consumers are feeding but in some cases it may be trivial, and in others interference may be more significant. As an extreme example, Székely and Bamberger (1992) studied the predation of chironomid larvae in lagoons by flocks of black-tailed godwits *Limosa limosa*, spotted redshanks *Tringa erythropus* and ruffs *Philomachus pugnax*. Exclosure experiments and observations of feeding rates showed that over thirteen days these birds removed 86.6% of the prey stock. As a result of this depletion, the intake rate from this food source declined by 32.7% for the black-tailed godwits, 14.3% for the spotted redshanks and 59.1% for the ruffs during the same period.

The objectives of this chapter are to describe a framework for studying depletion and give examples of the role depletion plays in determining the distribution of consumers. This framework will be extended in future chapters to consider the consequences for prey mortality, population size, predicting the consequences of habitat loss, and other issues.

3.2 Theoretical studies

A number of studies have examined the consequences of depletion in determining the distribution of consumers. In Royama's (1971) graphical model the consumers were free to move and feed in the patch with the highest density of prey. Once that patch was depleted so that the prey density equalled that in the next richest patch the consumers fed in both patches. This process continued with the consumers gradually feeding in an increasing range of patches.

Comins and Hassell (1979) compared the population stability of two host–parasitoid models: one in which the predators showed a simple aggregation in relation to prey density and one in which they continually shifted in relation to the depletion—as in Royama's model. Aggregation results in stability but the two models had similar levels of stability. Rohani *et al.* (1994) extended this approach with a spatial predator–prey model and incorporated realistic assumptions about the extent to which parasitoids will move between patches. They showed that this largely removed the stability. Bernstein *et al.* (1988, 1991*a*, 1991*b*) showed that the nature of prey mortality and the matching of predator and prey populations were greatly affected by both the learning of predators and travel costs.

Sutherland and Anderson (1993) provide a framework for describing the distribution of consumers between patches for species in which depletion is important and interference is not. The aim of this model is to describe the changes in the distribution of consumers over time and predict the population that can be sustained in a site.

This model considers a site with a number of patches that may differ in resource density. It assumes that there is no growth or reproduction of the resources. It also assumes that the resources do not move between patches. The number of resources taken, N_a, in time T in a patch with a resource density α is given by Holling's (1959) disc equation:

$$\frac{N_a}{T} = \frac{a'\alpha}{1 + a'\alpha\, T_h} \tag{3.1}$$

where T_h is the time it takes to handle each resource item and a' is the attack constant. This is explicitly defined by Nicholson and Bailey (1935) as the product of the area or volume of substrate searched in time T and the probability of detecting and obtaining those resource items within this area or volume.

The consumers initially feed in the patches with the highest density of resources. If the total area of these patches is f_M the initial rate of change in resource density within the patch is

$$\frac{d\alpha}{dt} = \frac{-a'P\alpha}{f_M (1 + a'T_h\alpha)} \tag{3.2}$$

where P is the number of consumers. As depletion proceeds, the resource density in the richest patch will be depleted to that in the next richest patch. Thus as time proceeds depletion will occur in a wider range of patches. After the maximum resource density M has been reduced by r resources the rate of depletion is

$$\frac{d\alpha}{dt} = \frac{-a'P\alpha}{\sum\limits_{j=0}^{r} f_{M-j}(1 + a'T_h\alpha)} \tag{3.3}$$

where the total area of patches with the resource density j is f_j. The time $t_{M.K}$ taken for the maximum resource density, M, to be depleted to a level K is then determined by rearranging and integrating eqn 3.3:

$$t_{M,K} = T_h \sum_{j=K+1}^{M} (j-K) f_j / P + 1/a' \sum_{j=K+1}^{M} (f_j/P) \ln(j/K). \tag{3.4}$$

The sum

$$\sum_{j=K+1}^{M} (j-K) f_j$$

is the total number of resources taken up to the instant when maximum resource density reaches K. Hence the first term in eqn 3.4 is the total handling time of resources up to this instant. The second term incorporates the influence of searching for resources.

Figure 3.1 shows, for a series of patches ranked according to initial resource density, how these patches are depleted over time to give the time necessary to deplete the prey down to a given level. The rate at which depletion takes place will be progressively slowed by two factors. Firstly, depletion extends over a greater number of patches (Fig. 3.1). Secondly, at lower resource densities the feeding rate will be reduced.

The intake rate of the consumers at different time intervals is then calculated by substituting values of the resource density (from eqn 3.4) into Holling's disc equation. As the patches are depleted the consumer's intake declines. Figure 3.2 shows the obvious fact that with more consumers the rate of depletion is greater and thus the intake rate is reduced at a faster rate.

The number of individuals that a location can sustain for a season of a given length can be determined from Fig 3.2. It seems reasonable to assume that individuals need a threshold intake I and individuals emigrate or die if their intake is less than this. This figure then shows the number of individuals that can be sustained for a season of a given length. For example, for a season of 150 days, 100 consumers would have excess resources, 250 could just survive, but 500 or 1000 would deplete their resources to such an extent that many would starve or need to emigrate.

The sustainable population can also be calculated analytically. As depletion proceeds, resource density will drop below the threshold resource density d_c which consumers need in order to obtain sufficient resource to stay alive—below this point consumers either starve or emigrate. For a given location to sustain P consumers over a season of length S it is necessary that, t_{M,d_c}, the time to deplete from the maximum resource density M to the threshold density d_c, must be less than the season, S.

Here t_{M,d_c}, is given by rearranging eqn 3.4:

$$t_{M,K} = \frac{1}{P} \{ T_h \sum_{j=d_c+1}^{M} (j-d_c) f_j + 1/a' \sum_{j=d_c+1}^{M} f_j \log(j/d_c) \} \tag{3.5}$$

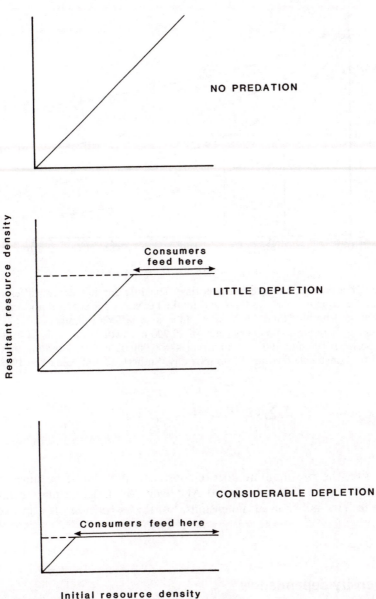

Fig. 3.1 The change in resource density with depletion after three time periods. The *x* axis shows the initial resource density in different patches ranked according to resource density. The consumers initially feed in the patch with the highest resource density, which they then deplete until the resource density is the same in the next highest patch and then deplete both. This process continues with the consumers gradually occupying an increasing range of patches.

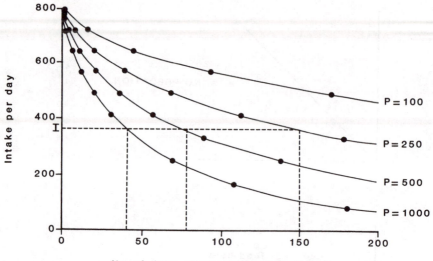

Fig. 3.2 The change in intake rate over time. The different lines are for different total numbers of consumers. I is the threshold intake necessary for survival and the dotted lines show the length of time this habitat can sustain different numbers of consumers. It is assumed there are ten patches, each of 10 000 m², with resource densities varying linearly between 100 and 1000 m⁻² and that $a' = 0.000\ 02$ m⁻² s⁻¹, handling time = 5 s, and the consumers feed for 12 hours a day (Sutherland and Anderson 1993).

so
$$P < \frac{T_h \sum_{j=d_c}^{M} (j - d_c)f_j + 1/a' \sum_{j=d_c}^{M} f_j \log(j/d_c)}{S}.$$
(3.6)

This gives the maximum number of consumers that can be sustained within a site for a season (Sutherland and Anderson 1993). This equation initially appears to have imbalanced dimensions, but this is because j is a measure of area.

3.3 Density dependence

As depletion proceeds in the absence of productivity, intake rate inevitably declines over time (Fig. 3.2). Goss-Custard (1980) calculated from field studies that a decline in prey density of 25–45% would result in a reduction in intake rate of 8–23% for redshanks and 5–18% for oystercatchers.

Obviously, the rate of depletion depends upon the number of consumers. Comparing the lines in Fig. 3.2 shows how intake rate will decline with the number of consumers. Thus at high population sizes, a higher proportion of individuals may starve.

Two pioneering field studies of depletion were carried out by Gibb (1958, 1960) and by Murton *et al.* (1966). Gibb (1958) showed that at the beginning of the winter the number of eucosmine moths *Ernarmonia conicolana* varies greatly between pine *Pinus sylvestris* cones, but as a result of predation by blue tits *Parus caeruleus* and coal tits *Parus ater* the numbers of moths were equally distributed by the end of the winter (Fig. 3.3). This is exactly in the manner outlined by the process shown in Fig. 3.1.

Coal tits were more likely to survive through those winters in which there was a high biomass of invertebrates in the foliage of the Scots pines (Fig. 3.4) (Gibb 1960). In three winters the birds took between 54% and 60% of the pupae. The combination of high depletion and a strong relationship between resource density and survival makes it very likely that depletion in this system results in density-dependent mortality. At a simplistic level, it is clear that twice the population could not be sustained in the same manner.

The study of Murton *et al.* (1964, 1966) showed that woodpigeons initially congregated in the fields with the highest clover densities and that depletion was greatest in these fields. They then moved to fields which had lower initial clover densities. The minimum size of the winter woodpigeon population was determined by the lowest level to which the clover density was reduced. The percentage of birds with food in their guts, the body weight of pigeons, and the overwinter decline in the pigeon population depended upon the abundance of clover in relation to the numbers of pigeons. As in the coal tits, it thus seems that depletion results in density-dependent overwinter mortality.

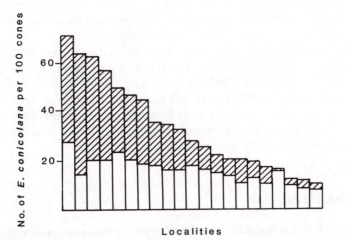

Fig. 3.3 Numbers of *E. conicolana* larvae per 100 cones before and after predation by tits. Each column represents one locality and they are arranged in descending order of the density of larvae. (From Gibb 1958.)

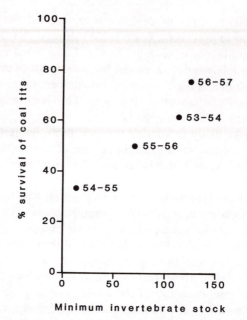

Fig. 3.4 The relationship between coal tit survival and the minimal inverte-brate density in Scots pine foliage. Each point is a different winter. (Redrawn from Gibb 1960.)

3.4 Buffer effect

The buffer effect is the phenomenon that at low numbers in a site, individuals collect in certain patches but as the population increases they occupy a greater range of patches (see Section 1.5). The same result is expected from the theory outlined here. As shown in Fig. 3.1, over time, consumers occupy an increasing range of patches as depletion equalizes the resource density.

The buffer effect can also arise from interference (see Section 1.5) or territoriality. A major difference is that if the buffer effect is caused by interference or territoriality then the process is immediately reversible: with a decline in population, individuals will concentrate in the preferred sites. By contrast, a buffer effect caused purely by depletion is irreversible if there is no productivity. If the consumer population declines, those remaining will not concentrate in the initially preferred patches.

3.5 Productivity

The approach so far has just considered resource depletion under the assumption that no productivity has taken place. In many cases the main period of productivity may differ from the period of depletion. For example, many migrant birds wintering in temperate areas feed on fruit or seeds whose productivity over the winter may be negligible. By contrast, the nectar in plants may be heavily depleted by nectivores soon after dawn but may be replenished by the next morning. Most systems are probably between these extremes. For example, many other migrant birds wintering in temperate areas feed on vegetation or intertidal invertebrates which may grow during the winter especially on warm days but most of the productivity takes place during the summer. If productivity can be measured it can be incorporated in the depletion models outlined here. For example, in Chapter 10, I describe models of geese feeding on grass in which the grass productivity depends on temperature.

One example of a natural system in which productivity occurs at a similar time and rate as depletion is the armoured catfish *Ancistrus spinosus*, which feeds by scraping algae from the substrata (Power 1974). Sixteen pools were compared in a stream in Panama and, not surprisingly, the algae grew faster in sunny pools than dark ones. The pools with the highest productivity also had the most fish. Depletion was important as exclosure experiments showed that reducing catfish densities resulted in higher standing crops of algae with higher productivity (Power 1990). Although the algal growth rate varied tenfold between pools, the calculated standing crop only varied twofold, suggesting that the fish distributed themselves so the depletion rates and algal growth rates were balanced out. In accordance with this, the growth rates and survival rates of the fish did not vary with algal biomass.

This balance between fish density and algal productivity was achieved by the movement of fish. During a November flood a very large sunny pool was created and a moderately sunny pool was partly destroyed. By January the fish had redistributed themselves such that the fish density again balanced the algal productivity. Similarly, in two experimental removals of fish the fish densities were restored within 9 and 17 days.

3.6 Combining interference and depletion

Interference and depletion have usually been studied in isolation although in most cases both will apply. Depletion will always be present although in some cases interference could be much more important in determining distribution. The difference is that with depletion the prey density within patches is declining and thus decisions as to where to feed will vary over time. The ideal free distribution with interference assumes no depletion and is best thought of as a snapshot at the given instant (Lessells 1995). When combining them it is necessary to determine the ideal free distribution with interference, then assess the depletion, and then calculate the new distribution. These two alternating processes are thus determined repeatedly.

The combination of interference and depletion may be modelled by dividing time into a series of intervals (Sutherland and Dolman 1994). For each time interval, the ideal free distribution can be determined for the interference and variation in competitive ability using eqn 2.3. The depletion in each patch can be calculated, thus giving the prey densities for the next time interval. This approach thus combines the depletion models of this chapter with the interference models of Chapter 2.

Figure 3.5 shows the aggregative response for a given value of interference, variation in competitive ability, and depletion rate. Each of the graphs in this figure explores the consequences of altering one of these parameters: (a) shows that at higher levels of interference the distribution is more even between patches; (b) shows the consequences of altering the extent of variance in competitive ability—as the variance is increased the distribution is more even and with very high degrees of variance there may even be more consumers in the patches of intermediate resource density; and (c) shows that at high levels of depletion the distribution is more even between patches as the depletion will reduce the differences in resource density. Chapter 2 gives models of interference alone and the results are similar to those of (a) and (b); however, interference without depletion is only likely to occur for short periods of time.

3.7 Continuous input

Although in most natural situations resource will be replenished at a slow rate, numerous experiments have considered cases where the resource is added at a continuously high rate. Thus although this has attracted considerable theoretical and experimental interest, it has few implications for the real world. One experiment, which works well as an undergraduate practical, is to provide bread at constant rates to two patches where a flock of mallard *Anas platyrhynchos* can feed (Harper 1982). As the ratio of rewards changes between the two patches, the ratio of the number of mallard changes in unison. Similar results were obtained during Milinski's (1979) study of sticklebacks *Gasterosteus aculeatus* when *Daphnia* were dropped into a fish tank at two points at different rates.

The concept of continual input may apply to ambush predators and to some foraging animals, such as fish competing for prey drifting downstream, but in reality such natural situations seem scarce. It is, however, readily testable and this allows us to see whether we understand the underlying behaviour in experimental situations. Continual input is important in considering males competing for arriving females such as dungflies *Scatophaga stercoraria* on cowpats (Parker 1970,1978) or birds on leks (see Chapter 6).

In continual input, the resource available to one individual may be taken by another. This is rapid depletion. We can express the rapid depletion in terms of m, the interference constant, although I stress it is not actually interference. The value of m, should theoretically equal one for continual input. The reason for this is illustrated in Fig. 3.6; increasing the number of competitors by tenfold or one hundredfold will obviously result in the mean reward declining to a tenth or a hundredth. Plotting the relationship between intake and consumer density on log scales must result in a slope of one if all the prey are taken.

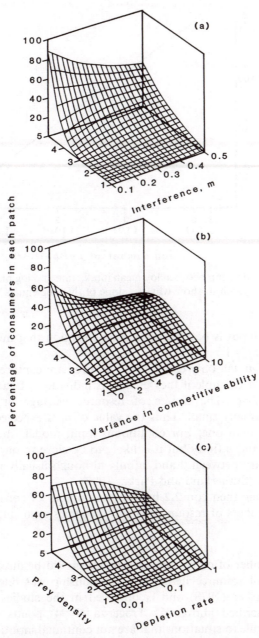

Fig. 3.5 The aggregative numerical response. This shows how the number of consumers in each of 20 patches varies: (a) with the degree of interference, m (all individuals are of equal competitive ability); (b) with the extent of competitive ability (interference is constant, $m = 0.2$); and (c) with the rate at which depletion proceeds (interference is constant, m = 0.2, variance = 4). Competitive ability is normally distributed over 21 phenotypes spanning ± 3 standard deviations around a mean of 10. Simulations consider 1000 consumers, feeding for 10 h per day for 150 days, in a site of five patches each of 20 hectares whose initial resource ranges from one to five items per square metre. Handling time = 20 s, quest constant, $Q = 0.001$ m^2 s$_{-1}$. (From Sutherland and Dolman 1994.)

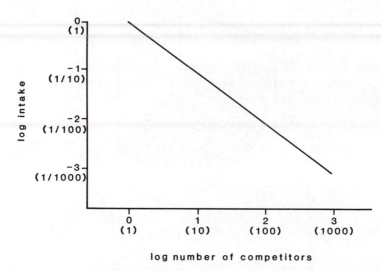

Fig. 3.6 The relationship between log mean intake rate and log number of consumers in a patch to show why the slope of this line equals one.

Thus continual input is best thought of as depletion that mimics interference with a value of $m = 1$.

The value of m will only equal one if all prey are captured. If some of the items are lost, for example if they escape as individuals fight over them, then m can exceed one. The degree of resource wastage in continual input experiments is usually small. Thus the value of m is expected to equal, or be slightly greater, than one. For continuous input models the value of m can vary between 1 and infinity but it is likely to be close to one; for interference models it can vary between 0 and infinity although usually it will be below 1 (see Section 2.1) (Sutherland and Parker 1992).

If m equals one then eqn 2.2 linking the number of consumers P_i in each patch to the numbers of resource items N_i becomes even simpler:

$$P_i = c \, N_i \tag{3.7}$$

Thus the number of consumers in a patch should be directly proportional to the number of resource items. This relationship was referred to as input matching by Parker (1978) and is observed in the studies of mallard and sticklebacks described above. As Tregenza (1994) points out, many have misapplied this rule to situations that are not continual input. There does seem to be a widespread belief that a universal prediction of the ideal free distribution is that the number of consumers in a patch should be proportional to the resource density. This prediction only applies to continual input.

Lessells (1995) provides another way of looking at continual input. At any given time consumers can distribute themselves according to the standing crop in each patch and the level of interference caused by other consumers. The standing crop is determined by the balance between the input rate and the consumption rate—which in turn is dependent upon the consumer density. Using such an approach, she shows that whether or not prey are consumed immediately has no impact on the expected distribution of consumers. However, if there is prey wastage then this may reduce the standing crop within the patches and hence affect the distributions of consumers.

Table 3.1 summarizes the results of experiments to test continual input models. In most cases the ideal free distribution appears to provide a reasonable, but by no means perfect, fit to the observed distribution. The expectation is that the intakes will be equal in each patch and the number of

Table 3.1. Tests of the continual input model of the ideal free distribution

Species	Result	Agreement with ideal free model	Reference
Guppy *Poecilia reticulata*	Intake somewhat higher in better patch	Yes, but poorer patch overused	Abrahams (1989)
Zebra fish *Brachydanio rerio*	Average intake same in both sites at low densities but average intake higher in better patch with high densities of fish	Yes at low densities; no at high	Gillis and Kramer (1987)
Cichlid fish *Aequidens curviceps*	Average intake same in two sites	Yes, but consistent individual differences	Godin and Keenleyside (1984)
Mallard *Anas platyrhynchos*	Average intake same in two sites	Yes, but consistent individual differences	Harper (1982)
Starlings *Sturnus vulgaris*	Overuse better patch	Contradicts model	Inman (1990)
Stickleback *Gasterosteus aculeatus*	Average intake same in two sites	Yes, but consistent individual differences	Milinski (1979, 1984, 1986)
Goldfish *Carassius auratus*	Intake somewhat higher in better patch	Yes, but consistent differences in success. Better patch overused	Sutherland *et al.* (1988)
Guppies *Poecilia reticulata*	Large differences in density between patches with identical treatments	Contradicts model	Talbot and Kramer (1986)

consumers will be proportioned to input rate. In most studies there are more individuals than expected in the low-input patch and these have a lower intake than in the high-input patch.

Individual differences in competitive ability have been shown in many studies of continuous input (Table 3.1) and these are best incorporated as relative differences in intercepts. For example, one individual may move faster than another and thus obtain more resource. In this case, the intake rate would depend upon the speed of an individual relative to the speed of others. This can be incorporated by:

$$a'_{(s,i)} = R_{(s,i)} \, Q^{-m} \qquad (3.8)$$

where, as before, R is the mean competitive ability of all individuals in a patch divided by the individual's competitive ability. This model produces a range of stable solutions (Sutherland and Parker 1992) as illustrated in Fig. 3.7. Sutherland and Parker (1985) pointed out that case E mimics the ideal free distribution and suggested that this case is expected most frequently by chance. Thus the apparent fit to the ideal free distribution shown by many studies in Table 3.1 may be an artefact. Houston and McNamara (1988) used statistical mechanics to show that this suggestion is not exactly right and on average we might expect slightly more individuals on the poorer patch than from the ideal free distribution.

This model of the ideal free distribution with unequal competitors (see Figure 3.7) was tested using goldfish *Carassius auratus* given a choice of two patches (Fig. 3.8) by Sutherland *et al.* (1988). There was a tenfold difference in intake between individuals. The rank order of intake rates was calculated and this rank was also consistent between the two patches; individuals with a high intake in the low-input patch also had a high intake in the high-input patch. The theoretical expectation of the model illustrated in Fig. 3.7 is that on those occasions when only a few individuals occupy a patch they will then be the better competitors. This experiment showed that when large numbers of fish were present in a patch they were of low mean rank, but when few were present they were of higher rank. A further theoretical expectation illustrated in Fig. 3.7 is that the good and poor competitors will have a similar distribution. In accordance with this there was no correlation between the competitive rank of the different goldfish and the time spent in the best patch. This thus differs from the examples listed in Table 2.1.

Many of the studies show fewer individuals in the better patches than expected from the ideal free distribution (Abrahams 1986; Kennedy and Gray 1993; Sutherland *et al.* 1988). Abrahams (1986) suggested that perceptual constraints may be important as individuals may find it hard to perceive differences in patch quality. The consequences of this are that the poorer patches are likely to be used more and better patches less often than expected from the ideal free distribution. A review of all published studies of continual input (Sutherland *et al.* 1988) showed that there is greater deviation from the

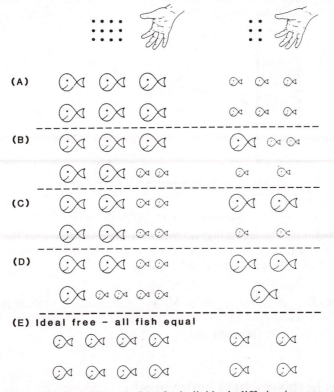

Fig. 3.7 The range of possible solutions for individuals differing in competitive ability. For this example food is added at rates of 6 and 12 items per unit time in the two sites. In A,B,C and D the fish differ in their competitive ability such that each of the big fish always capture twice as much as each of the six smaller fish. Note that (i) in none of the cases could any individual gain by moving; (ii) the sums of competitive abilities always equals 12 : 6; and (iii) the number of individuals and hence the mean intake in each site differ considerably between the different solutions. (From Sutherland and Parker 1985.)

ideal free distribution as the ratio of inputs increases. This is consistent with the idea that conceptual constraints limit the ability to choose between patches.

Kennedy and Gray (1993) suggested that the cost of travel between patches can explain the overuse of poor patches. Åström (1994) criticises this model and shows that the costs of travel should have no effect on the equilibrium distribution. It does, however, seem likely that an equilibrium will be reached more quickly if it is easy to move between, and sample, the patches.

Another reason for the overuse of poor patches was suggested to me by Manfred Milinski. Many of the continuous input experiments suffer from pseudoreplication as the same individuals are used repeatedly and often even for different input ratios. As individuals may use past experience in deciding which patch to use (Milinski 1994), this may produce a systematic bias and

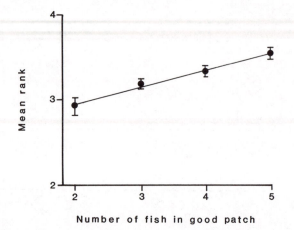

Fig. 3.8 Mean rank of individual goldfish (± 1 SE) plotted against the number of individuals in the high input site. ANOVA $F = 28.52$, $p < 0.001$). (From Sutherland *et al.* 1988; copyright Springer-Verlag, used with permission.)

move the equilibrium towards unity. Milinski (1979, 1984, 1986) is particularly careful in only using each fish in one trial which may explain why his results show the closest fit to the input ratio.

3.8 Does herbivory help herbivores?

I have so far made the assumption that as consumers deplete resources there will fewer resources available. However, for herbivores this may not always be true.

It is likely that many vertebrate grazers prefer patches with some interme- diate biomass over those with a higher or lower biomass. At low biomass there is obviously less food, but many species such as wigeon *Anas penelope* (see Chapter 10) and dark-bellied brent geese *Branta bernicla bernicla* also avoid longer swards (Vickery and Sutherland 1992). The reasons for this are unclear but longer swards probably consist of older leaves which are likely to be less digestible and contain less protein (Holmes 1989). Longer swards may require more energy to walk through and predators may be harder to see.

If a sward is shorter than the preferred height then grazing will make the patch less attractive to the consumer. However, if the sward is taller than that most preferred then grazing may make the sward more attractive. Hence herbivores may aggregate in patches that their presence continues to maintain (Fryxell 1991; Fryxell *et al.* 1988). Sheep selectively graze the best patches in lightly stocked pastures and the other patches deteriorate as they become older and less attractive. Over time the sheep thus become increasingly aggregated (Arnold 1964). Sea plantain *Plantago maritima* has increased protein levels following regrowth after being grazed by barnacle geese *Branta leucopsis*. It has even been suggested that the flocks return to the sward with a regular periodicity to take advantage of the regrowth (Ydenberg and Prins 1981). Brent geese feeding on the grass *Puccinellia maritima* also use patches of salt-marsh on a cyclical basis (Rowcliffe *et al.*, in press) with a periodicity that varies between 7 and 20 days; the shorter periodicities are for areas of salt-marsh with a higher proportion of *Puccinellia*. By the time the geese return the tillers are regrowing but have not fully regrown.

McNaughton (1984) showed that herbivores feeding in groups on the Serengeti plains have a higher intake than those feeding solitarily which he attributed to the grazers improving the sward. Westoby (1985) argued that this is rarely the case and referred to a large literature showing that in grazing trials in temperate grasslands weight gain of animals decreased with stocking rates. Hobbs and Swift (1988) suggested that the cause of the discrepancy between the two papers was due to the nature of the different ecosystems of the two studies. They suggested that feeding in grazed patches may be beneficial in habitats with high forage biomass such as that studied by McNaughton but detrimental to those habitats with low forage biomass such as that studied by Westoby.

An improvement of feeding conditions as a result of depletion makes modelling depletion less straightforward. One solution is to quantify the relationship between preference and biomass as shown in Fig. 10.2 for wigeon. It is straightforward to incorporate this into a model. As wigeon graze a high biomass patch it becomes more suitable until it reaches the height they prefer, after that any further depletion reduces suitability.

3.9 Switching between habitats

It is easiest to consider depletion as affecting the choice of patches within a habitat—as in the examples of the armoured catfish on algae, coal tits on pine cones, or woodpigeons on clover. It may, however, also be applied to the depletion within one habitat causing a switch to another habitat. Dark-bellied brent geese *Branta bernicla bernicla* in North Norfolk, as in many other sites in Britain, undergo a series of habitat shifts during the winter. In this section I will show how a detailed study of depletion improves our understanding of this switching.

As the geese arrive in late September they feed on intertidal algal beds, they then switch to areas of salt-marsh, and then in late October or November fly inland to feed on agricultural land. In the spring they return to the salt-marsh. It has been suggested that these habitat shifts may be related to resource depletion in coastal areas (Charman 1979; Summers *et al.* 1993).

The length of time the geese spend on the algal bed is determined by depletion and storm-induced losses. Exclosure experiments show that there is considerable depletion of algae in the autumn by goose grazing (Fig. 3.9). As

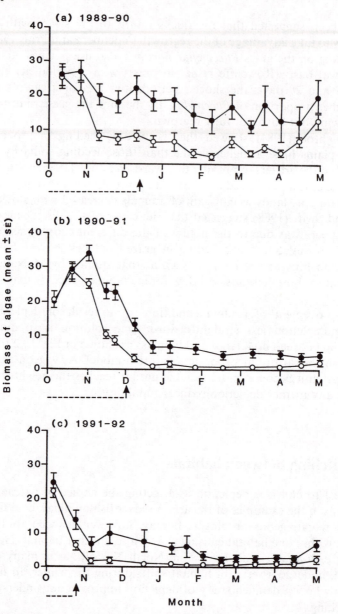

Fig. 3.9 Seasonal changes in biomass of the algae *Entromorpha and Ulvae* outside (o) and inside (●) enclosures in three winters. The arrow under the *x* axis indicates the point at which brent geese abandon the algal beds (Vickery *et al.* in press).

depletion proceeds, the geese spend a higher proportion of the time feeding and they walk at a faster rate. Yet, despite this greater effort there is a decline in intake rate, when measured as defecation rate (Fig. 3.10) (Vickery *et al.* in

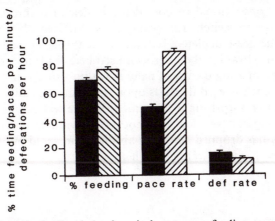

Fig. 3.10 Changes in feeding behaviour in brent geese feeding on algae as depletion proceeds. In each case the left bar shows the data for early in the season and the right the data for later in the season. (From Vickery *et al.*, in press.)

press). This reduced intake rate presumably precipitates their switch from the algal beds to the salt-marsh. In addition to the goose grazing, there is also loss of algae that correlates well with the strength of storms from northerly directions (Rowcliffe 1994). In 1991 much of the algae was swept away by autumn storms and the geese switched a month earlier (see Fig. 3.9).

It seems that the movement from the salt-marsh to the agricultural land is similarly caused by a decline in intake through the depletion and mortality of salt-marsh species. The decline of annual plants such as glasswort *Salicornia europea* was largely due to senescence but exclosure experiments showed that the grass *Puccinellia maritima* was significantly depleted by the geese (Summers *et al.* 1993). Once again, as depletion, senescence, and plant mortality proceeded, the geese fed for a greater proportion of the time, searching faster, yet had a reduced intake rate. This presumably explains why they switch from the salt-marsh to feed inland.

The world population of dark-bellied brent geese has increased dramatically over the last few decades as a result of legislation in Europe reducing shooting mortality (Ebbinge 1991). It seems inevitable that with a higher population, depletion will now proceed at a higher rate; in accordance with this, there have been dramatic changes in the timing of habitat shifts. In the 1950s the population in the north Norfolk study site was 245–450, and the birds switched from algae to salt-marsh in late February; in the early 1990s the population had increased to 4500 and they switched in October or November. Similarly, in the 1950s no birds fed inland, whilst birds now spend over four months inland. It is now possible to use the approaches of this chapter to predict the extent that further population increases will result in further increases in inland feeding (Chapter 10).

It has been conventional to consider the idea of a threshold biomass at which birds have to switch [for example see Madsen (1988) and Summers (1990)]. Once the geese deplete a habitat below this level they then move to a different habitat. However the approach used·here leads to a slightly different perspective. If comparing the use of patches within a site, as in Fig. 3.1, whether and when one patch is used depends upon the relative quality of this patch to others. It would be surprising if the same principle was not used in the choice of habitats. I suggest we view this not as individuals moving because the resource supply has dropped below an absolute threshold but because there is the possibility of higher rewards elsewhere.

3.10 Inequalities in consumers

In the approach outlined above it is assumed that all consumers are equal in their ability to find and handle resource. There is, however, evidence for differences between individuals. In this section I will review studies showing that individuals differ in competitive ability and argue that this has consequences for density-dependent starvation.

Most of the papers published on inequalities in the feeding behaviour of individual consumers consider age differences (Davies and Green 1976; Draulans 1987; Dunn 1972; East 1988; Espin *et al.* 1983; Grieg-Smith 1985; Groves 1978; Norton-Griffiths 1968; Recher and Recher 1969) but feeding rate may also vary with factors such as morphology (Ehlinger 1990) or sex (Smith 1975).

Following Orians' (1969a) classic study of brown pelicans *Pelecanus occidentallis*, numerous studies have shown that adult birds feed at a higher rate than juveniles. The more skill that is required to tackle prey, the less successful are the juveniles. Thus juvenile laughing gulls *Larus atricilla* are much less successful than adults when plunge diving for fish, slightly less successful when aerial dipping, and equally successful when catching offal in the air (Burger and Gochfield 1983). Juvenile sandwich terns *Sterna sandivicensis* are less successful than adults when diving for fish in deep water but as successful when diving from low heights into shallow water (Dunn 1972).

The age differences in feeding behaviour have been expressed in different ways: as differences in percentage success (Orians 1969a), handling time (Davies and Green 1976), or the ability to find resource (Draulans 1987). The relative importance of these processes was determined for moorhens *Gallinula chloropus* in a study comparing the behaviour of adults and juveniles (Sutherland *et al.* 1986). The walking rate was the same for adults and juveniles. Juveniles had both a longer handling time and a lower encounter rate, which resulted in a feeding rate that was 53% that of the adults. If the juveniles had the same handling time as adults they would feed at 59% of the adult rate, while if they had the same encounter rate they would feed at 83% of the adult feeding rate.

Older individuals may not always feed more efficiently. Very old red deer *Cervus elaphus* have worn cheek teeth which are poor at chewing tougher grasses and thus are the first to starve in times of low resource quality (Lowe 1969).

If all individuals are exactly identical then the expectation of the depletion model, as shown in Fig. 3.2, is the silly result that at the maximum density of consumers all individuals obtain sufficient resource to stay alive but with one more consumer they all starve! This seems deeply unrealistic although there is evidence for severe density-dependent mortality in soay sheep *Ovis aries* whose numbers increase until the vegetation is severely depleted and there is then mass starvation (Grenfell *et al.* 1992).

One reason why this expectation of all or nothing mortality does not usually occur in the real world may be that individuals differ in their feeding efficiency. As the resources are depleted, some individuals will starve or emigrate while others can survive on the same resources. I will consider other reasons in Chapter 8.

3.11 Summary

Depletion, the removal of resources by consumers, is an important process in determining intake and changes in distribution of consumers over time. The theoretical expectation of the ideal free distribution is that predators should reduce prey densities to a constant level in all patches. The theory can be used to determine the number of consumers a site can sustain.

Depletion may have important consequences for habitat switching and the buffer effect. Individuals differ in their foraging ability and thus may differ in the point at which they starve. Combining interference and depletion, makes it possible to describe the expected distribution between patches in terms of interference, depletion, and variation in competitive ability.

One example of rapid depletion is continual input in which resources are regularly replenished. This has attracted many experimental studies. Although the results are generally as expected from theory there is a consistent pattern that more individuals visit the poorer patch than predicted. A likely explanation for this is that individuals are limited in their ability to sample and perceive the best patches.

4

Prey availability

4.1 Introduction

Animals often appear to live amongst a superabundance of food. For example, for many thrushes, frogs, or damselfish there must often be as many prey in a square metre of substrate as an individual requires in a day. Herbivores are similarly often surrounded by abundant vegetation. Individuals may, however, starve amongst such apparent abundance and the reason, of course, is that not all of the food is available. Availability is thus a very important factor, yet it has received little attention as it is often hard to quantify.

The approaches used in the previous chapters assume that all the prey are equally available. There is, however, considerable evidence that this is not so. The aim of this chapter is to review the evidence for the role of variation in prey availability and consider the consequences of this for sustainable population size, functional and aggregative responses, and the pattern of prey mortality.

4.2 Sources of variation in availability

Many studies have shown that predators often select certain age classes. For example, large falcons feeding on dunlin in the Banc d'Arguin, Mauritania, caught a disproportionate number of juveniles and especially those that were thin, averaging some 26% below the mean weight (Bijlsma 1990). Wolves *Canis lupus* feeding on caribou *Rangifer tarandus*, dall sheep *Ovis dalli*, white-tailed deer *Odocoileus virginianus*, or moose *Alces alces* tend to select either fawns or old individuals (Curio 1976). It is thought that younger individuals are taken more often because they have less stamina in fleeing whilst old individuals may be susceptible due to increased disabilities. Cheetahs *Acinonyx jubatus* select male Thompson's gazelles *Gazella thompsoni* which tend to be less vigilant, and more isolated from others (Fitzgibbon 1990).

Location is also important. Those limpets *Notoacmea persona* and *Collisella digitalis* on steep surfaces beyond the reach of oystercatchers are much less likely to be eaten (Frank 1982). Studies of sanderling *Calidris alba* feeding on crustaceans (Myers *et al.* 1980), oystercatchers feeding on *Scrobicularia plana* (Wanink and Zwarts 1985) and curlew *Numenius arquata* feeding on the clam *Mya arenaria* (Zwarts and Wanink 1984) all show that the risk of predation is lower for prey deeper in the mud (Figure 4.1).

In many cases prey may be inaccessible for most of the time but susceptible to predation for short periods. Lugworms *Arenaria marina* are only vulnerable to bar-tailed godwits *Limosa lapponica* when they back up their burrows to defecate, and they are most likely to do this when the sand is wet (Smith 1975).

Availability may change over time and with environmental factors. The depths at which worms burrow is strongly correlated with the temperature—

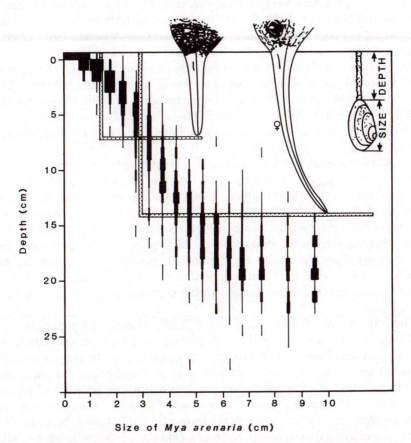

Fig. 4.1 The relationship between size and depth for the clam *Mya arenaria* and the range that can be taken by oystercatchers (left) and female curlews (right). This shows the number of clams of each size class occurring at each depth. The bars show the range of clams that oystercatchers and curlew can extract given that clams below a certain size are uneconomic to extract, and the birds are constrained by their bill length in the depth they can extract clams. (From Zwarts and Wanink 1984.)

when the upper surface of the mud is frozen very large worms may be as deep as 40–50 cm (Zwarts and Wanink 1993). Goss-Custard (1969) showed that below 4°C the rate at which redshank capture *Corophium* declines. The bivalve *Macoma balthica* buries much deeper in mid winter and thus is less accessible to wading birds (Reading and McGrorty 1978; Zwarts and Wanink 1993).

Little egrets *Egretta garzetta* in the Camargue feed largely on mosquito fish *Gambusia affinis* (Kersten *et al.* 1991). In the early morning, when macrophytes have depleted the oxygen from vegetated areas, the mosquito fish are forced to aggregate in open pools, where egrets can feed upon them easily. Soon after sunrise the oxygen level increases in the vegetated sites and the fish return and so avoid predation.

Individual prey may vary in their defences. Oystercatchers that specialize upon hammering through the shells of mussels select the thin-shelled individuals and leave those with thicker shells (Cayford and Goss-Custard 1990; Durell *et al.* 1984; Meire and Ervynck 1986; Sutherland and Ens 1987).

As Sinclair (1975) points out, most vegetation is inedible. The chemical composition of plants may vary between or within individuals and this is likely to be of considerable importance to herbivores. Meadow voles *Microtus pennsylvanicus* select reed canary grass *Phalaris arundinacea* leaves with low alkaloid content and eat little of those with high levels (Kendall and Sherwood 1975).

Parasites may make some individuals more susceptable to predation. Red grouse *Lagopus lagopus* killed by predators tended to have higher loads of the caecal nematode *Trichostrongylus tenuis* than did those birds that were shot by hunters (Hudson *et al.* 1992). Incubating females treated with an oral anthelminthic were less likely to be found by dogs trained to locate birds by scent. This suggested that females with a large parasite load emit more scent and thus are more vulnerable to mammalian predation. However, in some situations, prey with high parasite loads may be rejected. For example, oystercatchers reject *Macoma balthica* with high trematode infections (Hulscher 1973, 1982).

The importance of such individual differences in the prey population in relation to variation in availability will depend upon the ease with which prey can be captured. Temple (1987) compared the quality of the prey captured by red-tailed hawks *Buteo jamaicensis* with average members of the population. Characters such as age, parasite load, fat mass, and the presence of physical defects were all scored. Eastern chipmunks *Tamias striatus,* which were relatively easy for hawks to capture, showed no differences between those captured and those that were not, cottontail rabbits *Sylvilagus floridanus,* which are less easy to capture, showed differences in 3 of the 14 parameters between those captured and those not, while grey squirrels *Sciurus carolinensis,* which are the hardest for hawks to capture, varied in 10 of the parameters.

4.3 Quantifying availability

Quantifying availability is not easy. The best attempts come from the various detailed studies of wading birds by Zwarts and his colleagues. As the location of the prey in the mud can be determined and as the ability to extract prey is constrained by the length of a wader's bill, it is possible to undertake studies of selectivity that would be impossible for other groups.

One complexity is that the different sources of variation in availability may not vary independently. Knot ignore *Macoma* less than 9 mm long but cannot swallow prey longer than 16 mm. Furthermore, prey are inaccessible when buried more than 20 to 30 mm below the surface and the larger *Macoma* bury deeper. Thus only a fraction of the *Macoma* will be both of suitable size and within reach (Zwarts *et al.* 1992). Similarly, larger clams *Mya arenaria* are more deeply buried (Fig. 4.1). Male curlews have a bill of 11–12 cm and thus are restricted to taking smaller clams than are females with 14–16 cm bills. Most females feed on clams while most males do not (Zwarts and Wanink 1984).

Zwarts and Wanink (1984) measured the size and depth of clams taken by one colour-ringed curlew (known as 20 yellow) with a bill length of 14.3 cm. This showed (Fig. 4.2) that 20 yellow was very good at finding the very few large clams that were much shallower than expected for their size.

As well as influencing the number of prey found, prey availability has other effects on the intake rate. Not only are large individuals deeper, but for a given length, the heavier individuals were deeper in the mud (Zwarts and Wanink 1991). For a given size, deeper individuals were 50% heavier in the bivalve *Scrobicularia plana*, 40% heavier in the worm *Nereis diversicolor*, 25% heavier in the bivalve *Macoma balthica*, and 20% heavier in both the clam *Mya arenaria* and the cockle. As a consequence of this pattern, wading birds selecting prey near the surface may be consistently having to take the lighter prey.

Detailed studies of prey behaviour may help to understand the profitability of different habitats. Esselink and Zwarts (1989) showed that worms *Nereis diversicolor* not only buried deeper as they grew larger but that worms of similar size buried on average 2 to 3 cm deeper in sand than in mud. However, in experiments, the burrowing depths of worms of similar size and weight proved to be the same in sand as in mud. The explanation for the difference between habitats was that, for a given length, those worms in better body condition were able to maintain deeper burrows, and that the body condition of worms living in sand was 20–35% higher so they occurred at a greater depth. This helps contribute towards the greater profitability of mud-flats than sand-flats to waders.

4.4 Theory

There have been previous studies considering differences between individual hosts in susceptibility to parasites or parasitoids. Bailey (1962) considered the

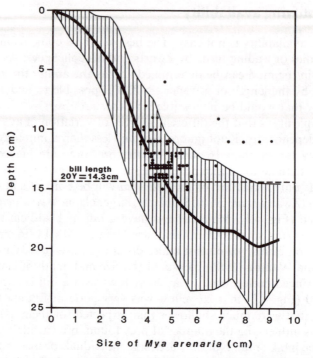

Fig. 4.2 The size range and depths of clams taken by an individual colour-banded curlew '20-yellow' with a bill length of 14.3 cm (dotted line). The hatched area shows the depth range at which 95% of the clams of each length are living and the bold line gives the mean depth of each size. The points show the clams actually extracted. (From Zwarts and Wanink 1984.)

consequences of some individuals in a prey population having higher predation risks, Smith and Mead (1974) considered the consequences of there being both vulnerable and invulnerable age classes, and Hassell and Anderson (1984) considered variation in host susceptibility within host–parasitoid systems.

In order to examine the consequences of prey availability on the selection of patches by consumers and their intake rate, I created a model. I considered a series of eight patches each of 10 000 m² which differ in prey density. The prey density varies linearly between these patches so that it is eight times higher in the richest patch (280 m⁻²) than in the poorest. I then assumed that the prey vary in the ease with which they can be detected. This was incorporated by assuming a log distribution of prey availabilities (the values of attack constants a' for the seven classes were 0.01, 0.0046, 0.0021, 0.001, 0.000 46, 0.000 21, and 0.0001). In each patch a seventh of the prey population was assigned to each availability.

The intake rate for each patch was then calculated using Holling's (1959) disk equation. To make it clear that handling time is not responsible for the shape of the functional response I have assumed handling time is zero. I assume that, as in the depletion models of Chapter 3, the 100 consumers all feed in the patch where their intakes are highest. Depletion occurs as the individuals feed in the richer sites and so reduce prey density. There is no interference incorporated in this model.

Unlike the depletion model of Chapter 2, the pattern of depletion depends also upon the availability of the different classes within each patch. Figure 4.3 shows the obvious phenomenon that over time the accessible prey are removed, leaving many of the less accessible prey. Exactly the same process is observed with humans selecting pistachio nuts at parties—the proportion that are hard to open increases with time.

The exploitation of prey is not uniform. In the patches of high density a greater proportion of the accessible prey are removed (Fig. 4.3). Unlike depletion of identical prey in which the prey are reduced to a constant density in each patch (see Chapter 3), in this model individuals are using patches of low resource density not because all other patches have been depleted to this level but because of the depletion of the most accessible prey. For the situation shown in this figure, most of the prey are still present yet the depletion of accessible prey still has considerable consequences for the distribution of consumers.

4.5 Functional and aggregative responses

If all prey are identical, then in the absence of interference, the expectation is that lower resource density patches are ignored and higher density patches are all depleted to a constant level. With a constant resource density in all occupied patches the expectation, if the consumers are identical, is that the consumer density should be equal and their intake should be equal. Thus the theoretically expected functional response and the expected aggregative response are both a single dot. This is because all the patches that are exploited should theoretically be depleted to exactly the same density and the density of consumers and their intake rates should thus be the same in all these patches. As depletion proceeds, more patches will be used but they will only be used when the prey density and intake rate equals that in the other exploited patches. It is, of course, naive to expect such a result, but it is necessary to explain the field observation that the intake rate is reasonably constant over a wide range of resource densities (Fig. 2.7) and why consumers feed over a range of patches varying in resource density (Fig. 2.6) rather than collecting in the patch with the most resources.

The model outlined here can provide what I believe is probably a major explanation of the shape of functional and aggregative responses. With variation in availability, the predation in the richer patches results in depletion

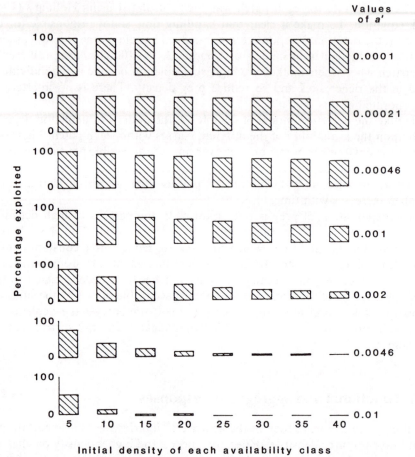

Fig. 4.3 The pattern of depletion affecting eight patches (arranged along the *x*-axis) differing in resource density. In each patch the resource is equally divided into seven different availability classes (values of a' shown down right hand side). This graph shows the percentage unexploited of each availability class in each patch after 100 days. See text for details and other parameters.

of the more available prey. Individuals thus feed in the patches of lower prey density because they contain a greater proportion of the more available prey (Fig. 4.4).

This analysis provides an explanation for the shape of functional responses such as those shown in Chapter 2. The standard explanation for the shape of functional response curves is that the plateau is a consequence of the handling time or satiation. The intake is independent of resource density when most of the time is spent handling prey rather than searching for it. I argue in Section 2.1 that this explanation is often insufficient. Figure 4.4. shows that an asymptotic functional response arises from this model of depletion and

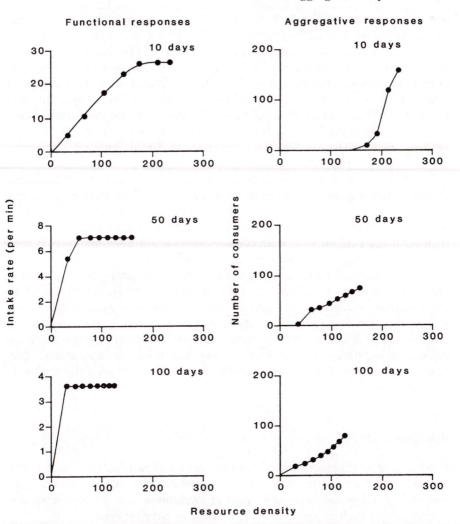

Fig. 4.4 The functional response (left) and aggregative response (right) shown for different periods as depletion proceeds from the model outlined in Fig. 4.3. Note that the *y*-axis changes as depletion proceeds.

variation in prey availability. With Holling's disk equation the asymptote should be the reciprocal of the handling time but with zero handling time in this model the asymptote should be infinite! Intake rate is the same over the range of patches in which the consumers feed, despite the absence of handling time. Similarly, satiation is not included in this model and cannot be the explanation of the lack of increase in intake with resource density. The explanation of this is that the amount of exploitable prey is depleted to a constant level in each patch. Thus this may well explain why the intake may be constant over a range of patches even though the prey density varies.

Individuals may starve even when there are abundant resources. Suppose that an intake of 4 items per minute is necessary to stay alive. From the functional response shown in Fig. 4.4, after 10 days it seems that any patch with at least 40 resources per m² will provide sufficient intake and thus each patch may be depleted to this level. However, after 100 days, the intake in each patch has dropped to below 4 items per minute although the resources are up to three times higher than the apparent threshold. Simple measures of resource abundance may thus often be insufficient.

The theoretical expectation if all prey are equally available is that the higher prey densities will then be depleted to an even level. With differences in prey availability this need not be the case. This then has consequences for the pattern of prey mortality. If there is no variation in prey availability the predation is strongly density dependent. With considerable variation the mortality is more evenly distributed and less strongly density dependent.

I have assumed for simplicity that at each resource density there is an equal proportion of each availability. In reality there will often be complex relationships between availability and density. For example, at a given tidal height in areas where they are abundant, mussels tend to form clumps that may contain more small individuals. Furthermore at different tidal heights there are considerable differences in mussel density, shell thickness, and standardized flesh content (Goss-Custard *et al.* 1993). All of these factors influence intake (Hulscher 1982).

4.6 Summary

Food may seem superabundant to us although some individual consumers may be starving. This discrepancy may be a consequence of variation in prey availability. There are large differences in availability between individual prey due to factors such as age, sex, morphology, or parasite load.

Variation in prey availability complicates any study that includes the resource population. This may help explain why consumers do not just congregate in the patches of highest densities but feed in patches of a range of prey densities, why the intake rate is relatively constant between patches that vary considerably in prey density, and the nature of density-dependent prey mortality.

5

Prey populations

5.1 Introduction

The previous chapters have considered the behaviour and distribution of the consumers. The aim of this chapter is to explore the consequences of the distribution and intake rate of consumers for the prey or resource population. In particular, how does the mortality rate vary with the prey density?

5.2 Theoretical pattern of prey mortality

In earlier chapters I outlined models that describe the distribution and intake of consumers feeding in a range of patches varying in prey density. This approach can be used to explore the manner in which the spatial variation in prey mortality will vary with the behaviour of the consumers.

With this approach, the theoretically expected pattern of density dependence depends upon the depletion rate, the strength of interference, and the variation in competitive ability. If only depletion takes place, the expectation is that the consumers initially aggregate in the patch with the highest prey density and feed there until it is depleted to the density in the next richest patch. Thus they deplete the prey in each patch to a constant level and as a result the prey mortality will be density dependent.

The pattern of prey mortality will also vary with the extent of consumer interference (Lessells 1985; Sutherland 1983). In the model described here, interference was incorporated using eqn 2.2. Figure 5.1 shows the consequences of varying the value of m, the interference constant. If interference is negligible then, as described earlier in this section, the expectation is that all individuals feed in the richest patches and the prey mortality will be density dependent. With greater values of m, individuals will distribute themselves over a greater range of patches and mortality will then be less-strongly density dependent. If m equals one then the expectation is that the distribution of consumers will match that of the prey and the resulting mortality will be independent of the prey density. If interference, m, exceeds one the ratio of predators to prey will

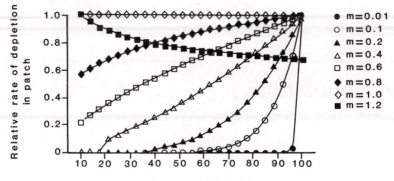

Fig. 5.1 The nature of density-dependent mortality for different levels of interference expressed relative to the rate of depletion in the highest patch. This is a modification of the model used elsewhere (see Figs. 3.5, 8.1, and 8.2) to describe the distribution of individuals in a patchy environment. This considers 5000 consumers in an area of 1000 h. The patches are divided into 10 different prey availabilities varying linearly between 10 and 100 items m⁻². Each prey density occupied a tenth of the total area. The intake rate is determined by Holling's (1959) disc equation. The handling time = 1 s, the Quest constant, Q, = 0.005. Consumers forage for 10 hours per day for 150 days. (From P.M. Dolman and W.J. Sutherland, unpublished.)

be highest in the poorer patches and the prey mortality will thus be inversely density dependent.

Incorporating individual differences in competitive ability makes this more complex. A wider range of patches will be used as the poorer competitors will occupy the poorer patches and prey mortality will thus be spread more evenly between patches. If there is considerable variation in competitive ability, the numbers of consumers may be highest at intermediate resource densities as only the individuals of very high competitive ability occupy the richest patches. This results in the highest prey mortality at the intermediate prey densities (Fig. 5.2).

The expected pattern of prey mortality will also depend upon the total number of consumers in the site (Fig. 5.3). At low populations, all individuals will collect in the patches of highest prey density and the prey mortality will be strongly density dependent. As the total consumer population increases, the interference will increase in the richer patches so that more individuals will use the poorer patches. The prey mortality may then be less strongly related to density and may even show a domed response with highest values at intermediate levels.

The pattern of predation is likely to change the variance in prey density. If, as seems often to be the case, the prey mortality will be density dependent, then this will reduce the variation in prey density. If, however, it is inversely density dependent, then it will increase the variance. If the mortality is independent of density then it will have no effect as it will deplete each patch equally.

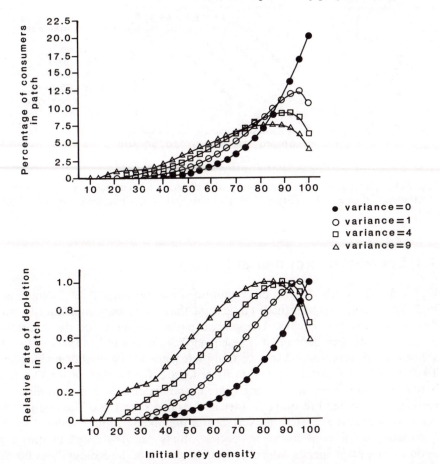

Fig. 5.2 The nature of density-dependent mortality for different degrees of variation in competitive ability. This was incorporated by a normal distribution of competitive abilities across 21 phenotypes around a mean of 10. Same model as Fig. 5.1. (From P.M. Dolman and W.J. Sutherland, unpublished.)

As described in Chapter 4, variation in availability will affect the pattern of mortality. This is easiest to consider by assuming there is no interference. If all prey are equally available then theoretically the prey should be depleted to a constant level in each patch. However, if variation in availability is incorporated, then consumers may start feeding on a lower density patch as a higher proportion of the prey in the higher density patches are inaccessible (see Sections 4.4 and 4.5).

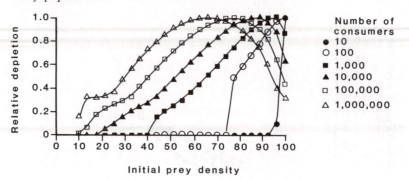

Fig. 5.3 The pattern of prey mortality with different total populations of consumers. Same model as Fig. 5.1. (From P.M. Dolman and W.J. Sutherland, unpublished.)

5.3 Examples of prey mortality

Figure 5.4 shows the studies of vertebrate consumers that P.M. Dolman and W.J. Sutherland (unpublished) could find that relate prey mortality to prey density in different patches. The general pattern seems to be that mortality tends to be density dependent. Lessells (1985) did a similar analysis for the studies of invertebrates and showed that these tended to be density independent. Two possible explanations for this difference are, firstly, that vertebrates have better sampling ability and so aggregate in the better sites and, secondly, that vertebrates may tend to move further in the daily selection of feeding patches yet studies of invertebrates and vertebrates may be on similar scales.

Whether a given herbivorous species inflicts density-dependent mortality upon a given plant species will depend on the details of the ecology. Van Eerden (1984) measured grazing impact as dry weight of *Salicornia europea* seeds (which are contained within the plant) removed by wigeon and barnacle geese. Although the total grazing increased with seed density, the actual predation is inversely density dependent; a maximum of 70% was removed at a biomass of 20–50 g compared with 50% removed at the highest biomass (110–120 g). By contrast, in studies of teal feeding on seeds of the grass *Agrostis stolonifera* the predation is density dependent. The small size of the seeds leads to exploitation only occurring once a minimum seed density is exceeded.

Any relationship between the distributions of consumers and resources will become more complex if the prey are mobile and especially if the prey move to avoid the predator. Thus, if the prey move randomly we may expect no relationship between predators and prey, while if the prey move much more rapidly and evade the predator we may expect a negative relationship between them (Sih 1984). This is also likely to be dependent upon the scale of observation chosen. Studies of Atlantic cod *Gadus morhua* feeding on capelin *Mallotus villosus* (Rose and Leggett 1990) show that at scales of 4–10 km the densities were positively correlated while at scales of 2–3 km they were

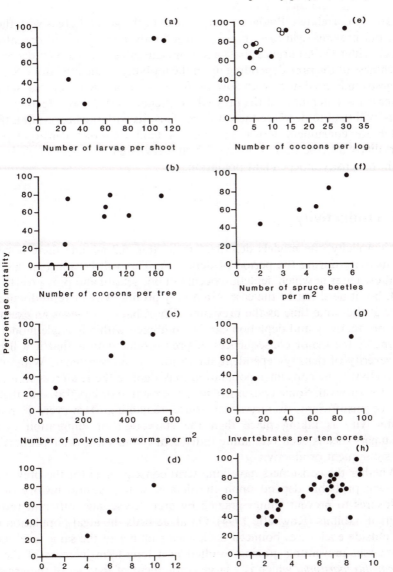

Fig. 5.4 Examples of density-dependent prey mortality. (a) pardalot *Pardalotus ornatus* and yellow-tailed thornbill *Acanthiza chrysorrhoa* on late instar nymphs of the pysillid *Cardiaspina albitextura* (Clark 1964); (b) white eye *Zosterops lateralis* on codling moth *Cydia pomonella* cocoons (Wearing 1975); (c) short-billed dowitcher *Limnodromus grisens* on the tubicolors polychaete *Clymenella torguata* (Schneider 1992); (d) turnstone *Arenaria interpres* on the sea urchin *Echinometa lucuntor* (Schneider 1985); (e) blue tit *Parus caeruleus* on codling moth *Cydia pomonella* cocoons (Wearing 1975) (Symbols show results from two different experiments); (f) three-toed woodpecker *Picoides tridactylus* and other woodpecker on spruce bark beetles *Dendroctonus rufipennis* larvae (Knight 1958); (g) various waders on intertidal invertebrates (Schneider 1978); (h) American pine siskin *Carduelis pinus* on winter moth *Operophtera brumata* larvae Roland *et al.* (1986). (From P.M. Dolman and W.J. Sutherland, unpublished.)

negatively correlated. Predators may collect in the general areas with the most prey but prey may, in turn, make localized movement to avoid predators.

Schneider (1992) argues that predators can either increase or decrease the patchiness of the prey depending upon the mobility of the predators. Schneider suggests that predators with low mobility, such as snails or sea stars, will increase the patchiness of the prey whilst predators that move readily, such as birds, will tend to decrease the variance. This fits with the observation that the vertebrate consumers shown in Fig. 5.4 tend to inflict density-dependent mortality while the studies of invertebrates reviewed by Lessells (1985) tended to show density independent predation.

5.4 Productivity

For simplicity, the models described in Section 5.2 assume there is no prey productivity during the period of depletion. This will be approximately true for those cases in which depletion occurs in one season and prey growth in the next, but it need not be the case; often prey productivity or reproduction take place at the same time as the prey mortality. Chapter 10 gives an example of how productivity and depletion can be combined within a single model.

One of the major consequences of prey productivity is that it will reduce the severity of density-dependent starvation by the consumer. Without prey productivity, the consumer population may deplete the resources to a critical level for survival. Some consumers may then starve or emigrate but there will still be insufficient resources for those remaining. However, if resource productivity is taking place then the starvation or emigration of some consumers means each remaining individual obtains a greater proportion of any subsequent productivity.

Whether the consumers have long-term consequences for the prey population will probably depend upon the details of the natural history of each. Exclosures to prevent winter grazing by brent geese gave different results in different habitats (Rowcliffe 1994). On algae beds the algal population inside and outside enclosures bounced back each summer to the same level. On the salt-marsh, preventing grazing resulted in a long-term increase of the grass *Puccinellia maritima*, which is a large component of the diet of the geese.

5.5 Summary

The behaviour of predators can be used to predict the spatial nature of prey loss. Depending upon the values of interference, depletion, and variation in competitive ability, the prey mortality may be density dependent, density independent, inversely density dependent, or with greatest mortality at intermediate densities. Reviewing the published studies suggests that vertebrate predation is usually density dependent.

6

Territories

6.1 Introduction

In the previous chapters it has been assumed that individuals are relatively free to move between patches. However, in many species this is clearly untrue as some individuals defend territories that exclude others. In this chapter, I will consider how territory defence influences the distribution of individuals and consider the consequences for density dependence. In some species territories may be defended by groups.

Territorial behaviour is usually most pronounced in the breeding season and this is the situation I consider here. However the same approach can also be applied to territoriality in the range of conditions under which it occurs. For example, the majority of Neotropical migrant passerines are probably territorial in their wintering grounds and have high site tenacity between years (Fischer 1981; Rappole and Warner 1980); the redshank is even territorial in winter (Selman and Goss-Custard 1988) but not on the breeding grounds (Hale 1980); some species, such as pied flycatchers *Fidecula hypoleuca*, may hold territories on migration (Bibby and Green 1980); and grey herons *Ardea cinerea* and white-fronted bee eaters *Merops bullockoides* breed in colonies but also defend seperate feeding territories (Hegner and Emlen 1987; Marion 1989).

The aim of this chapter is to describe how the choice and size of territories has consequences for distributions, density dependence, co-operative breeding, and delayed maturation.

6.2 Ideal despotic distribution

The ideal despotic distribution of Fretwell and Lucas (1970) describes the distribution expected when the first individuals to arrive are able to gain sole access to parts of a patch. As illustrated in Fig. 1.3, each patch may be considered as consisting of a range of territories which may be ranked in quality with the first individual to arrive occupying the highest-quality territory. Thus, for each settling individual, the crucial factor is the suitability of the best remaining territory within each patch. As each individual pre-empts the occupancy by another, Pulliam and Danielson (1991) suggested that this is termed the ideal pre-emptive distribution.

In an ideal world, individuals will start by occupying the patch with the highest-quality territories. Other patches will only be occupied when the highest-quality territory in that patch is of the same quality as the highest-quality remaining territory in the best patch. Thus, although the average quality of territories may differ between patches, the quality of best unoccupied available territories should theoretically be the same in all patches (Pulliam and Danielson 1991).

6.3 Settlement patterns

Territories that result in the highest breeding success are often the first to be occupied in the spring (Bensch and Hasselquist 1991; Brooke 1979; Lanyon and Thompson 1986). Various studies have also shown that the most-productive territories are occupied more frequently (Andrén 1990; Baeyens 1981; Møller 1982).

The relationship between settlement and habitat quality may depend upon the scale examined. In territorial white-footed mice *Peromyscus leucopus* living in nest boxes there is a clear fitness advantage for females occupying forest relative to edge and fence row habitats and this is reflected in their habitat choice (Morris 1989). However, within habitats there was no evidence that they preferentially used those boxes in which the mice had a higher reproductive success (Morris 1991).

Resource abundance will be only one of a number of factors in determining the suitability of territories. Møller (1991) studied seven species of birds in a number of small woods varying between 0.01 and 3.61 hectares. For five of these species (great tit *Parus major*, blackbird *Turdus merula*, yellowhammer *Emberiza citrinella*, chaffinch *Fringilla coelebs*, and magpie *Pica pica*) the brood size at fledging was greater in larger woods (Fig. 6.1). The differences in brood size were not due to differences in clutch size or resource abundance (Møller 1987) but due to the higher predation risk in the smaller woods (Fig. 6.2) as shown in other studies (Andrén and Angelstam 1988; Wilcove 1985). Møller (1988) placed plasticine eggs in old nests and examined these for peck marks which confirmed that the attack rate was greater in smaller woods. Three species (great tit, house sparrow *Passer domesticus*, and swallow *Hirundo rustica*) nested in protected sites and thus were not susceptible to predation, and as expected the brood size of swallows and house sparrows did not show any variation with wood size. Predation was not, however, the entire explanation for all species with the higher brood size in larger woods. Great tits also nest in protected sites yet this species showed an increase in brood size with wood size, and four of the species using unprotected nest sites still showed an

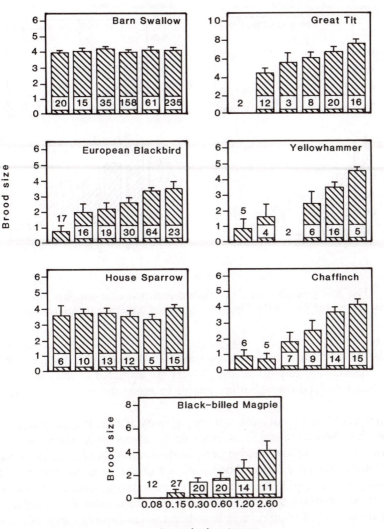

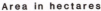

Fig. 6.1 The brood size at fledging in relation to wood area for seven species of passerines (Møller 1991). The numbers inside boxes are the number of broods.

increase in brood size with wood size, even after the clutches eaten by predators were excluded from the analysis.

It is likely that it is often difficult for individuals to assess territory quality. The abundance and behaviour of predators may play a large role in determining the likelihood of breeding success and yet this may be difficult to assess. Kentish plovers *Charadrius alexandrinus* in Hungary breed in alkaline grasslands and drained fishponds and switch back and forth between these habitats during the breeding season (Székely 1992). Due to the higher predation risk in the

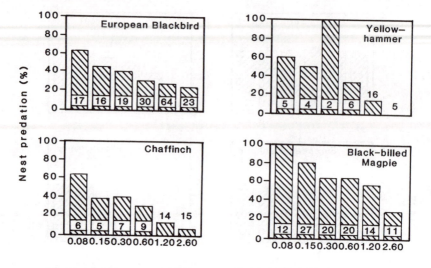

Fig. 6.2 The nest predation rate in relation to wood area for four open-nesting passerine species (Møller 1991). The numbers inside boxes are the number of nests.

fishponds the breeding success there is only half that of the grassland habitat. However, Székely suggested that the plovers are attracted to the fishponds as there is abundant food there, but are unable to assess the associated predation risk.

Another problem in assessing territory quality is that breeding territories will often be chosen before it is possible to assess prey abundance. The distribution of male yellow-headed blackbirds *Xanthocephalus xanthocephalus* was unrelated to the abundance of odonates, which are the major prey provided to the young (Orians and Wittenberger 1991). Females did settle on the marshes with the most odonates but their choice of male territories was unrelated to resource abundance. This may have been because females regularly fed outside the male territory or simply due to difficulties of sampling by the females. It has even been suggested that the reason why Great Reed Warblers *Acrocephalus arundinaceus* usually return to the same marsh is that birds can use the knowledge of the different territory qualities gained in the previous year (Bensch and Hasselquist 1991).

Other factors may also influence settlement patterns. In some territorial systems, such as those of many carnivores, inheritance of territories from the parent may be more important than settlement pattern (Lindström 1986). It has been suggested that the shape of the territory may be determined more by the ease of defence than by foraging considerations (Eason 1992).

I will discuss two approaches to considering the pattern of territory occupancy. In the first (Sections 6.4–6.7) I assume that territories are fixed in location and size. In the second (Section 6.8) I assume that the territory size can fluctuate according to the costs and benefits of defending different sized areas.

6.4 Theory of fixed territories

For the models described in Sections 6.4–6.7, I assume that territories are relatively fixed in position and size. I assume territory quality, expressed as average breeding success, may vary due to differences in resource availability, nest site suitability and predator abundance. This thus ignores factors such as variation in the number and quality of mates or in the probability of the adult surviving.

Figure 6.3(a) shows three examples of how territory quality may vary. In one case all territories are identical and each has the same average reproductive output while the other two cases show different degrees of variation in output.

If the ideal despotic distribution is strictly obeyed then the territories are occupied in the exact order of suitability. Thus at low densities only the best territories are occupied and the mean reproductive output is high whilst as the density increases poorer territories are occupied and the mean reproductive output is reduced. The resulting relationship between mean reproductive output and density is shown in Fig. 6.3(b) for the three different extents of territory variability. The density dependence is obviously stronger if there is considerable variation in territory quality.

The density dependence resulting from variation in territory quality assumes individuals have perfect knowledge of the quality of each territory and settle accordingly. Perfect knowledge is obviously implausible; as suggested for the yellow-headed blackbird and kentish plover described earlier, there may be a poor fit between resource abundance or predation risk and the settlement pattern. With imperfect knowledge density dependence will be weaker and with complete ignorance there is obviously no density dependence from this mechanism (Pulliam and Danielson 1991).

6.5 Floaters

Many studies have shown that there are individuals that are referred to as 'floaters' present on the breeding grounds that do not hold territories (Smith and Arcese 1989). For example, Mönkkönen's (1990) study of passerines in northern Finland shows that after experimental removal of territory holding individuals, the free territories were rapidly occupied by smaller-sized individuals.

As described earlier, the expectation of the ideal despotic distribution is that as population density increases, poorer quality territories are used. If all the unoccupied territories are poor quality then it may pay individuals not to breed at all but to wait until one of the better territories may become available due

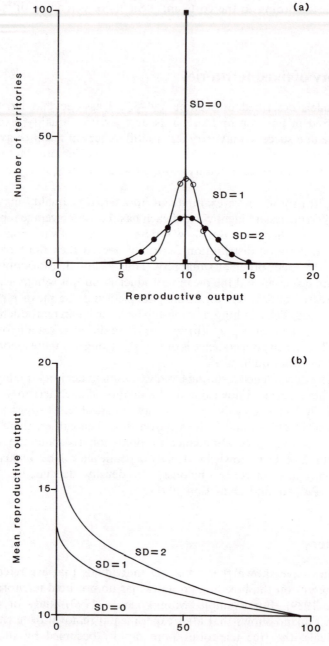

Fig. 6.3 The relationship between density-dependent reproductive output and variation in territory quality. (a) Three systems with different extents of variation in reproductive output among territories; in one, all territories are of equal quality, and in the other two the reproductive output per territory varies with 1 or 2 standard deviations. (b) The relationship between mean breeding output and population size for the three systems shown in (a): individuals occupy the best territory available so that poorer territories are only occupied as the population increases.

to the death of the occupant. Ens *et al.* (1992) emphasises the importance of this process and suggests modelling it as a queue (a line in North America) and I have adopted his term here. With this terminology floaters are examples of queueing individuals.

Queueing can be incorporated into the current model by considering the lifetime reproductive success obtained from occupying the best unoccupied territory compared with the lifetime reproductive success obtained from acting as a floater waiting to occupy an available territory.

For individuals that choose to breed in the best available territory the lifetime reproductive success (LRS) is the fecundity in this season plus the fecundity in all future seasons in which it is alive. Thus, with an age-independent survival rate, *S*, and an age-independent annual fecundity rate, *F*:

$$LRS = F + \sum_{n=1}^{\infty} S^n F \qquad (6.1)$$

For individuals that queue, the lifetime reproductive success depends not only upon whether the individual survives to breed in future years but whether the present occupant dies beforehand. The lifetime reproductive success can then be calculated as

$$LRS = \sum_{n=1}^{\infty} (S^n - S^{2n}) F \qquad (6.2)$$

I consider here the simple case of each floater waiting to occupy a given territory. Such a model can be readily modified to consider floaters that could occupy a number of territories.

Figure 6.4 shows the average reproductive lifespan for an individual occupying a vacant territory and for an individual that queues in an occupied territory. If the annual probability of survival is low then a queueing individual is unlikely to survive to breed. If survival is high then the relative difference between strategies is smaller. Of course if individuals are immortal then it never pays to queue!

If individuals adopt the strategy that results in the highest lifetime reproductive success, then it is possible to determine the expected degree of queuing for given combinations of survival rate, density, and variation in territory quality. As the density increases, a higher proportion of potential breeders queue rather than breed in the poorer quality territories as shown in Fig. 6.5(a). This was determined by comparing the lifetime reproductive success from the above equations for breeding and queuing for each additional individual. The individual adopts the strategy with the highest lifetime reproductive success. The variation in fecundity is incorporated in the model by varying the standard deviation in breeding output from each territory. Many studies have shown that younger individuals breed if the density is reduced, for example in ospreys *Pandion haliaetus* (Poole 1989) and sparrowhawks *Accipiter nisus* (Wyllie and Newton 1991).

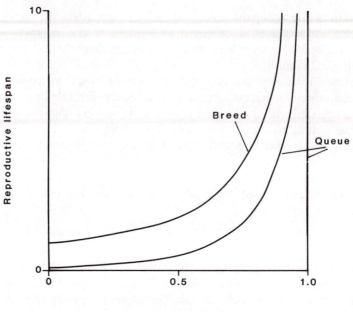

Fig. 6.4 The reproductive lifespan for individuals that breed or queue as derived from eqns 6.1 and 6.2.

Figure 6.5(b) illustrates how the percentage of individuals queuing increases with survival rate. With a low survival rate the queuing individual is relatively less likely to be alive when the territory owner dies. It follows from Figure 6.4 that if the current territory holder is immortal then it does not pay to queue.

Figure 6.5(c) shows how the percentage of the population that queues increases with the variation in habitat quality. If all territories are equal then it is always better to breed than queue if a territory is vacant, while if there is considerable variation in territory quality then it pays some individuals to queue. Thus the different components of Fig. 6.5 show that queueing is favoured if territories vary greatly in quality, if population density is high, and if survival is high.

6.6 An example of queueing

The clearest demonstration of the link between individual decision making, territoriality, and floaters is the study of Ens and colleagues of oystercatchers on the Fresian island of Schiermonnikoog. Ens *et al.* (1992, in press) shows that the birds occupy two different types of territory (Fig. 6.6). The strategy of some birds, known as residents, is to defend territories that overlap both

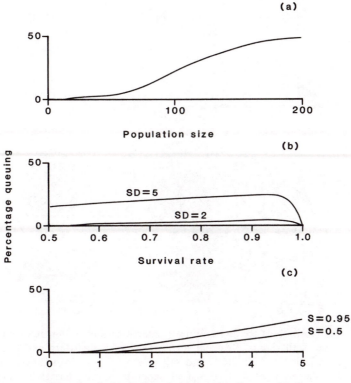

Fig. 6.5 The relationship between the percentage of individuals queuing and (a) population size (survival = 0.8, SD = 5), (b) annual survival rate, with two values of standard deviation in breeding output (population = 100), and (c) standard deviation in territory quality, expressed as breeding output, with two values of survival rate, *S* (population = 100).

the sea edge of the salt-marsh and the adjacent mud-flats. Residents nest on the salt-marsh and then walk with their chicks onto the mud-flats. These territories are highly compressed laterally which implies that the competition for space is intense. The strategy of other birds, known as leap-frogs, is to defend both an inland salt-marsh nesting territory and a feeding territory on the mud-flats. Leap-frogs cannot take their chicks to the mud-flats, as they are attacked if they enter the territories of residents, and thus have to fly the 200–1000 m between the feeding and nesting territories whilst carrying the food for the chicks. This behaviour is not unusual as many other coastal oyster-catcher populations consist of a mix of residents and leap-frogs.

The breeding success varied between years but, on average, the resident oystercatchers fledged between two and six times as many offspring as did the leap-frogs. Ens *et al.* (1992) estimated that to provide sufficient food the

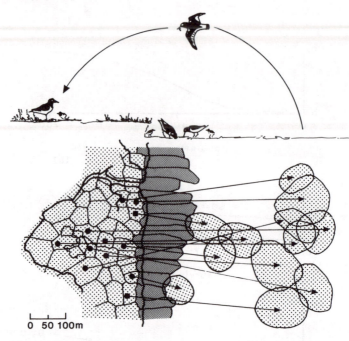

Fig. 6.6 The position of feeding and nesting territories of oystercatchers on Schier-monnikoog, the Netherlands. Resident territories are shaded dark and overlap both mud-flats (on which the birds feed) and salt-marsh (on which they nest). Leap-frog territories are shaded light and consist of seperate nesting territories and feeding territories. (From Ens *et al.* 1992.)

leap-frogs would have to fly over an hour each low tide yet no adult ever reached this level of effort—most leap-frog chicks simply starved to death.

In determining which birds adopted which strategy, Ens *et al.* (in press) excluded the following four hypotheses: that recruits discriminate badly, that leap-frogs were poorer quality, that the low reward of leap-frogs was balanced by low mortality, and that leap-frog territories were a stepping stone to better quality territories. The data supported the hypothesis that the gains from the enhanced reproductive success of residents is balanced by the longer delay in becoming a resident. A model was created (Ens *et al.*, in press) showing that if most individuals aimed to become residents then it paid to be a leap-frog and vice versa. From this model it was possible to determine the evolutionarily stable solution at which the benefits of the two strategies are equal.

As shown in Figure 6.5, at high population densities individuals should not breed in poor territories but queue instead. Ens *et al.* (in press) suggest that this is a major reason for deferred maturity in many species. For example, oystercatchers are capable of breeding when three years old yet many do not breed until a few years later.

6.7 Co-operative breeding

A very similar relationship between delayed breeding and territory quality to that shown by oystercatchers is shown in the co-operatively breeding Seychelles warbler *Acrocephalus seychellensis* (Komdeur 1992, 1993). The main difference is that queueing Seychelles warblers stay on their parents' territory and help raise the young. The birds in the better-quality territories (defined as having high foliage cover and high insect density) had a greater breeding success. Young birds from the poor-quality territories dispersed whilst those on the better territories tended to stay with their parents and act as helpers. Komdeur calculated that birds on the high- or medium-quality territories obtained, on average, a higher lifetime reproductive success from helping and queuing than from moving to a poorer-quality territory and breeding immediately.

Helpers would breed if the territory holders died. As part of a conservation programme, Seychelles warblers were moved from their sole location, the island of Cousin, to the nearby islands of Aride and Cousine. The 38 gaps created on Cousin by the removal of breeding birds were rapidly filled by helpers or one-year-old birds, sometimes within hours. The birds introduced to Aride and Cousine soon bred, but with the low initial population and plenty of unoccupied high-quality habitat there was a complete lack of co-operative breeding. Unlike on Cousin, the offspring from all pairs moved to create new territories in unoccupied high quality areas. As the population increased, and all the better habitat was occupied, the situation began to resemble that on Cousin with the first-year birds on good-quality territories staying and helping.

There seems to be a difference between species, such as oystercatchers and Seychelles warblers, in which young show deferred breeding and queuing for the better sites, and the passerines studied by Møller (1991) and Krebs (1971) which did not. In the oystercatchers and Seychelles warblers, individuals occupying the poorer territories never moved to the better territories but in great tits the experiments of Krebs (1971) showed they did. Why should this be? It is likely that floaters are quicker and better at exploiting empty territories than are individuals occupying poorer territories. Thus the presence of floaters in the population may exclude the possibility for others to both breed in a poor patch and then move if a better territory becomes available. Hence if the combination of values of survival rate, variation in territory quality, and density are such that some individuals are floaters (as in Fig. 6.5) then this may exclude the option of others both breeding and searching for better territories and thus result in even higher numbers of floaters.

6.8 Cost–benefit analysis

There is a long history of using cost–benefit analysis to consider territory size (Brown 1964). There are a number of benefits to possessing a large territory (Davies and Houston 1984) which have been shown to include an increased food supply (Jarman 1974; Snow and Snow 1984), reduced cannibalism by neighbours of eggs (Black 1971) or young (Ewald *et al.* 1980; Sherman 1981), attracting a female earlier in the season (Davis and O'Donald 1976), attracting more females (Verner 1964), or reduced predation (Dunn 1977; Krebs 1970). Costs are harder to quantify but are assumed to be related to the problems of defending against intruders.

The standard approach is to consider the costs and benefits increasing with territory area, and the optimal territory area occurs where the difference between these lines is greatest (Brown 1964; Davies and Houston 1984). The cost–benefit approach seems to give sensible results. This approach has also been useful in predicting when Hawaiian honeycreepers *Vestiaria coccinea* (Carpenter and MacMillen 1976) and pied wagtails *Motacilla alba* (Davies and Houston 1983) should defend territories.

In this section I extend this cost–benefit approach to consider the distribution of individuals differing in competitve ability in a site that consists of patches differing in quality. This is a game theory model so that the behaviour will depend upon the behaviour of others. For simplicity, I assume that the gain from a given territory is proportional to territory area multiplied by the patch quality (for example this may be thought of as the total amount of food present in the territory) but that the costs increase exponentially with territory area. The optimal territory size is then the area at which the benefits minus costs is greatest. This is determined by differentiating the slopes and setting them equal (Parker and Knowlton 1980).

Individuals differ in competitive ability which is likely to influence their territorial behaviour. I incorporate this by assuming that the costs of territory defence depend upon whether the individual is a good or poor competitor and use a similar approach to that used in Chapter 2 (Parker and Sutherland 1986; Sutherland and Parker 1985, 1992).

$$cost = k \, e^{-RA} \tag{6.3}$$

where k is a constant, A is territory area, and R is the relative competitive ability defined as the mean competitive ability of all other consumers in the patch divided by the individual's competitive ability. At high densities the total area of territories may exceed the area available, and then the value of k is progressively increased until this is no longer the case—the increased competition for space will lead to increased costs of territory defence with the result that either territories are compressed or some individuals go elsewhere or do not defend territories. With this formulation the defence costs are greater for a low-quality individual than a high-quality one, and for an individual of a given quality, the costs are greater if others in the patch are high-quality individuals than if they are low quality.

I assumed a small fixed benefit for those individuals that adopt the strategy of floating, due for example to reduced mortality, the possibility of extra pair copulations, or the ability to occupy rapidly vacant territories. Thus, under some circumstances, it pays individuals to be floaters rather than defend a territory. The distribution of competitors can then be described. As outlined in Chapter 3, the technique for modelling the evolutionarily stable strategy for individuals of each competitive ability is to distribute individuals of each phenotype in proportion to net benefit. This then produces the distribution at which all members of each phenotype obtain the highest net benefit taking into consideration the behaviour of all members of the same and other classes.

The optimal territory size is that area at which the net benefit is greatest. It initially seems likely that as good competitors have lower costs they should be expected to have larger territories. This is indeed the result obtained from the model when all patches are set to have equal value (Fig. 6.7). The combination of a larger territory and a better quality individual is expected to result in higher breeding success. This leads to the straightforward expectation of a relationship between quality and territory size and between territory size and reproductive success. In this section I will argue that these sensible expectations need not be correct when there is a range of patches.

Incorporating a range of patches differing in quality makes the outcome more complex. Figure 6.8 shows how territory size changes with population size. At small populations the optimal territory is largest in the better patches. As population size increases the increased competition in the better patches

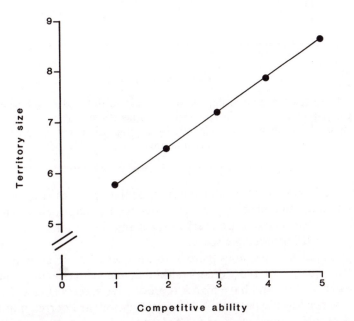

Fig. 6.7 The relationship between optimal territory size and competitive ability when patches are equal in quality. The site comprises 8 patches, each of 100 units in size; the population consists of equal numbers of each of 5 phenotypes whose competitive ability varies linearly between 0.8 and 1.2. Benefit for floating = 4.

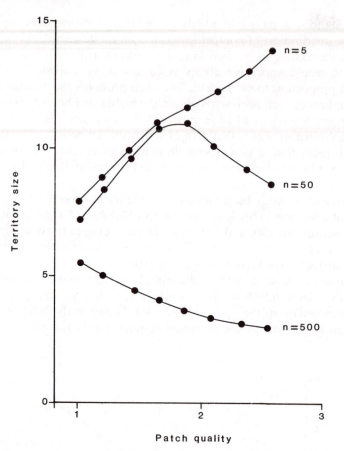

Fig. 6.8 The expected relationship between mean territory size and patch quality for different population sizes. The poorest has a benefit per unit area of 1 and this increases by 0.2 for each subsequent patch, so that the best patch has a benefit of 2.4 per unit area.

results in the largest territories in the intermediate patches. Furthermore, at high populations, the territory size decreases with patch quality. This shows that in the field we need not necessarily expect any simple relationship between territory size and reproductive success.

At large population densities there is no clear relationship between territory size and the quality of individuals—although the better-quality individuals occupy better patches and have higher reproductive success (Fig. 6.9). Thus in the field, we need not expect to find any straightforward correlation between territory size and individual quality. As a possible example, bluegill sunfish show a significant negative correlation between home range and body size (Fish and Savitz 1983), and it is suggested that larger, more successful fish occupy smaller more profitable home ranges.

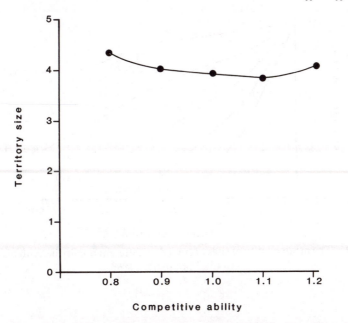

Fig. 6.9 The expected relationship between territory size and competitive ability when individuals are distributed between patches within a site when population size is high.

Schoener (1983*b*) points out that the predictions of territory size vary according to the assumptions of the model used. My object here is not to produce precise models of exactly how territories operate but to consider the general ways in which competition and the ideal despotic distribution affect territory size and to show that standard predictions may no longer be true when unequal individuals are distributed between a range of patches.

6.9 Buffer effect

The buffer effect (Brown 1969*a*, *b*) is that at low densities individuals tend to occupy the better patches while at higher densities a greater proportion occupy the poorer patches (see Section 1.5). As examples, the mean territory quality in a population of nuthatches *Sitta europea* decreased with increasing population density (Nilsson 1987), and red squirrel *Sciurus vulgaris* females occupying poorer territories raised fewer young and these poorer territories remained vacant at lower population densities (Wauters and Dhondt 1990).

The nature of the buffer effect resulting from the cost–benefit model just described can be explored. As the population size increases, poorer quality individuals increasingly move to the poorer patches (Fig. 6.10) such that the proportion in the better patches declines with total density (Fig. 6.11). Furthermore a higher proportion do not breed, but queue.

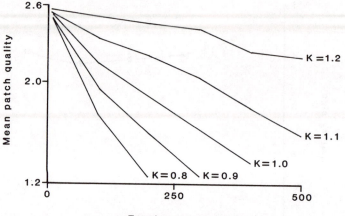

Fig. 6.10 The mean quality of sites occupied by different quality individuals in relation to total population size.

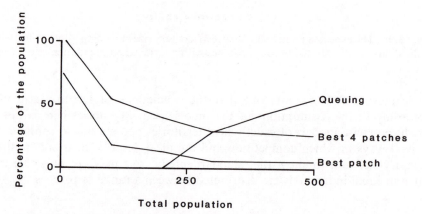

Fig. 6.11 The buffer effect: the expected fraction of individuals occurring within the better-quality patch in relation to the total population size. This is the same model as in Figs. 6.8, 6.9, and 6.10.

O'Connor (1986) used the British Trust for Ornithology's Common Bird Census to examine the changes in distribution of farmland birds in response to changes in overall density. The population of many species plummeted in the cold winter of 1962–63 and then recovered to previous levels over subsequent years. This provided the opportunity to see how distribution changed in relation to density. As the density of mistle thrushes *Turdus viscivorus* increased fourfold, the birds occupied a wider range of habitats. In the less frequently used habitats, eggs were laid later and were less likely to survive. In years when the national density of wrens *Troglodytes troglodytes* was low, the number of wrens on a given farm was correlated with the extent of hedgerow on that farm while at a high national density of wrens this correlation disappeared as they then settled on farms with few hedges.

O'Connor (1986) also explored the idea that use of less preferred habitats should be more variable than the use of preferred habitats (Fig. 6.12). The highest densities of dunnocks *Prunella modularis* occur at 7–9 m of hedgerow per ha; the variance in numbers was lower for the preferred habitat than for areas with either more or less hedgerow. Analysis of five other species showed that blackbirds had a similar optima, magpies were more stable at high densities of hedgerows, mistle thrushes were more stable on farms with less than 2% woodland, but wrens and chaffinches showed no clear patterns.

In the model presented here poorer competitors occupy poorer patches and are more likely to be floaters. The general pattern from field studies is that it is the young individuals that are floaters or helpers or who occupy the poorer territories. For four of the seven species studied by Møller (1991) and discussed in Section 6.3, a greater proportion of yearling birds occured in the smaller woods (Fig. 6.13), but this pattern was not shown for house sparrows or swallows. Krebs (1970) showed that it was the youngest great tits that occupied the poorer hedgerow territories and these moved to woodland territories if they became vacant. As well as age differences there may be other differences, for example, male pied flycatchers *Ficedula hypoleuca* in the preferred deciduous woods were larger and heavier than those in coniferous woods (Alatalo *et al.* 1985).

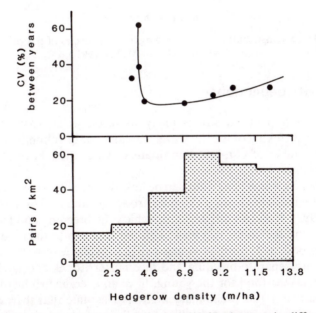

Fig. 6.12 The relationships between the length of hedgerows in different areas and both the coefficient of variance in annual numbers of dunnocks in each area and the median population density. (From O'Connor 1986.)

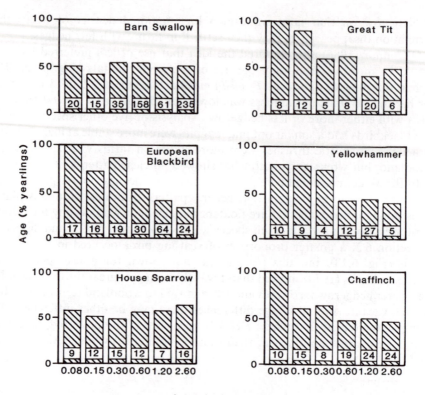

Fig. 6.13 The age distribution (as % yearlings) of six species of passerines in relation to wood area. (From Møller 1991.)

6.10 Density dependence

An increase in population size is likely to result in at least one of three possibilities: a shrinking of territory size, a greater use of suboptimal territories, or a greater number of non-breeding floaters. All of these have consequences for the breeding output

The manner by which behaviour can be linked to density-dependent breeding output can be illustrated using the data from the study of oystercatchers by Ens *et al.* (1992). Resident territories, in which the breeding and feeding areas are adjacent, and the young can walk with the adults onto the mud-flats, have an average productivity of 0.64 young per pair, while leap-frog territories, in which there are separate breeding and feeding territories and adults have to fly back with all the food for the young, have an average productivity of 0.17 young per pair. For this preliminary analysis I assume that there are a fixed number (25) of each territory type. The breeding rates per pair are then halved to give the number of young per adult. The lifetime reproductive success of each member of the queue for a resident or leap-frog territory can then be calculated. The first 25 individuals of each sex hold territories immediately and their lifetime reproductive success (*LRS*) equals

$$LRS = F + \sum_{n=1}^{\infty} S^n F \qquad (6.4)$$

where S is the survival rate and F is the mean number of young produced per breeding adult per year, which is much higher for resident territories than leap-frog territories. Further individuals have to wait before breeding in year B and their lifetime reproductive success equals

$$LRS = S^B \sum_{n=B}^{\infty} S^n F \qquad (6.5)$$

The delay before breeding depends upon the position in the queue and the mortality rate. The lifetime reproductive success of each individual in the queue for a leap-frog territory can be calculated in exactly the same manner but with a lower value of F.

The relationship between mean breeding success and density is determined by assuming each individual joins the territory or queue position with the highest *LRS*. The mean breeding output is then total breeding success of all breeders divided by the number of birds, including those that queue and so do not breed.

At low populations, only the resident territories are occupied and as a result the mean breeding output is high. At higher populations two processes result in a reduced mean breeding output: some individuals become leap-frogs and others queue. In theory the shape of the relationship depends upon the survival rate but in practice it seems that this has little effect.

Once the nature of density-dependent breeding output is known, then this can be used to consider the consequences for equilibrium population size. The number of fledglings produced can be converted into the number of birds surviving to return to the breeding grounds by multiplying by the survival rate from fledgling to adulthood. The lower curve of Fig. 6.14 thus shows the predicted density-dependent rate at which adult oystercatchers are produced. As described in Chapter 1, population size can be considered as the interaction of the density-dependent nature of both mortality and breeding output. The mortality rates can then be introduced into this analysis to determine the equilibrium population size. Figure 6.14 thus shows the expected population size for the typical mortality rates of 0.05 (Hulscher 1989). As shown in Fig. 6.14, the consequence of increasing the mortality to 0.1 can have a considerable influence on population size. It is thus clear how this work be used to examine the consequences of a change in survival due to say shooting, pollution or habitat loss.

The aim of this model is to show how an understanding of behaviour can be used to derive the relationship between breeding output and oystercatcher density. In turn this can be used to consider the factors influencing population

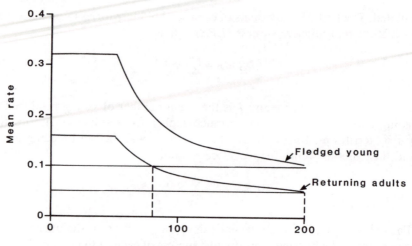

Fig. 6.14 The mean reproductive output per pair of oystercatchers in relation to the oystercatcher population assuming territory sizes are fixed. The upper curve shows the number of fledged young produced per adult. The lower curve incorporates survival to adulthood to give the number of returning adults per parent. The horizontal lines show mortality rates of 10% and 5%. The equilibrium population is the point at which the mortality rate crosses the lower curve.

size. This model is preliminary and other factors will also probably be important. For example, this model assumes all territories can be divided into two types while in reality there will be a range of habitat qualities.

It is conventional to consider habitat saturation as if there is a clear upper limit on the number of territories and indeed in some situations this may occur. However the models in this chapter show that detecting habitat saturation need not be straightforward. At high populations some individuals may become floaters rather than occupy poorer territories and an ecologist observing floating individuals and also observing that if a territory holder dies its territory becomes occupied may conclude that all the habitat is saturated. However, should the population increase further, individuals may settle in the poorer territories showing that the area was not truly saturated. Thus with this perspective there may not be a rigid upper limit.

Sinclair (1989) showed that the calculated density dependence in breeding parameters for vertebrates is usually very weak. However, whether individuals breed or not may well be more important than declines in, say, fecundity or egg production, but as floating individuals are difficult to monitor non-breeding may be much harder to detect.

Although it is difficult to measure, the extent of failing to breed seems likely to be one of the strongest contributions to density-dependent loss in breeding output. In song sparrows *Melospiza melodia*, both the proportion of individuals that were floaters and the probability of territory holders failing to find a mate were density dependent (Smith and Arcese 1989). Newton (1992) reviewed the experimental manipulations of territorial birds and of the 56 species studied there was evidence in 45 that individuals were excluded from breeding by the territorial behaviour of others.

A. Dobson (pers. commun.) reanalysed the data of Southern (1970) on tawny owls *Strix aluco* in Wytham Wood, England. As the population increased, mean fecundity declined. This was entirely due to occupancy of territories with lower breeding success—the fecundity in each of the territories did not change with density. The density-dependent decline in mean clutch size in blue tits and great tits has similarly been attributed to the increasing use of poorer patches (Dhondt *et al.* 1992).

Although at high densities some individuals will fail to obtain a territory and others will occupy poor areas, the consequences for the breeding output might not be as great as initially imagined. It is probable that the individuals occupying the territories are also the most able breeders while the floaters are likely to be juveniles and subdominants that would tend to have a lower breeding success anyway. Such differences in breeding ability will make the expected density dependence less strong.

In the model described here, I only consider the benefits of occupying territories for breeding success; but territories may be necessary for survival, for example, brown trout *Salmo trutta* populations are largely regulated by density-dependent survival in the juvenile stage (Elliott 1989, 1990, 1994). Territory size is dependent on fish size but not on the density of fish. Larger territories require greater effort to defend from intruders and the effort spent in territory defence increases markedly when there were large numbers of trout without territories. At high densities of non-territorial trout, territorial individuals spend much of each day in territorial defence. For the larger juveniles the cost is probably too great to defend a large territory and this may explain why medium-sized juveniles are most likely to survive. Territory defence can thus result in strong density-dependent mortality both through individuals failing to obtain territories and through individuals dying from the effort of excluding others.

Similarly, Grant and Kramer (1990) used published data to determine the relationship between the length of seven species of juvenile salmonids and the size of their territories. This was then used to predict for each species the maximum number of territories that could be fitted into a given habitat and thus could be used to determine the pattern of density dependence.

6.11 Summary

The ideal despotic distribution can be used to explain the pattern of settlement in territories. One approach is to assume that territory, location, and size are fixed. An alternative approach is to assume that territory size may vary according to the relative costs and benefits of defending it. Models incorporating a range of patches of different quality give different results from those considering only a single patch.

As population density increases, individuals are more likely to occupy poor-quality territories, smaller territories, and to fail to breed. All of these contribute to density-dependent loss in breeding output. It is then possible to explore the links between aspects of territorial behaviour and population size.

At high densities individuals may obtain a higher lifetime reproductive success by refraining from breeding in poor territories and waiting to occupy a better territory made vacant by the death of the occupant.

7

Mating systems and reproductive success

7.1 Introduction

In this chapter I will consider how ideal free theory can be used to consider mating systems and especially lekking behaviour. Leks are communal mating grounds in which males display and females visit to mate, gaining no resources from attending the lek—see Wiley (1991) for a review. The approaches used here are similar to those used in previous chapters, but I will restate some of the basic concepts described in previous chapters within a framework of mating systems. I will sketch out some of the ways in which mating systems can be studied using the ideal free distribution and describe lek mating in more detail.

7.2 Distributions of males and females

The great diversity of mating systems implies that it will be difficult to provide a simple framework to understand them all. Clutton-Brock (1989) suggested that much of the variation in mating systems could be explained by a combination of female dispersion, male dispersion, and the extent of parental care. Lott (1991) also relates the variation in social systems found in vertebrates to the spacing behaviour of individuals.

It is widely accepted that it is sensible to consider females distributed according to factors such as resources, predation, and the costs of social living, and then to consider the distribution of males as determined by the distribution of females (Fig. 7.1) (Davies 1991, 1992). This is expected on theoretical grounds. In species in which the females contribute more to the offspring, the male reproductive success is determined largely by the number of mates, while the female reproductive success is determined by the resources (Bateman 1948; Trivers 1972)

As examples of the relationships shown in Fig. 7.1, experimental studies of pied flycatchers suggest females settle in relation to resources rather than to male distribution (Alatalo *et al.* 1986; Slagsvold 1986); placing caged female voles *Clethrionomys rufocanus* in the field determined the location of free living

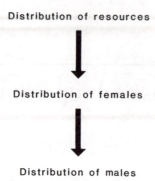

Fig. 7.1 The suggested relationship between resources and the distribution of females and of males.

males but caged males had no effect on the location of free living females (Ims 1988); and the distribution of female dunnocks *Prunella vulgaris* is determined largely by the food supply while males are distributed in relation to females (Davies 1992).

The distribution of females can be considered in relation to resources, such as food density and nesting sites, and also to predator abundance, and in relation to competition from other females due to interference, depletion, or territorial exclusion (Bradbury *et al.* 1986). Exactly the same approach that is used for determining females in relation to resources can then be used for considering the distribution of males in relation to that of females: male distribution will depend upon the location of females and the location of other males. The possible mating systems depend upon the degree of overlap in the distribution of females and males.

A number of studies have examined whether the distribution of males searching for females corresponds with that expected from the ideal free distribution. Davies and Halliday (1979) show that the distribution of searching male toads *Bufo bufo* is consistent with each male obtaining equal numbers of females. There are also two studies of invertebrates, of dungflies (Parker 1970) and tiger blue butterflies *Tarucus theophrastus* (Courtney and Parker 1985), in which the distribution of males searching for females corresponds with the ideal free distribution.

7.3 Interference

As described in Chapter 3, interference is the decline in the rate at which resources are obtained as the number of competitors increases. This is often described by the equation

$$a'_i = Q \, P_i^{-m} \tag{7.1}$$

where a'_i is the searching efficiency of a consumer in patch i, P_i is the consumer density in patch i, Q is the Quest constant (the value of a' achieved by an individual feeding alone), and m is the interference coefficient (Hassell and Varley 1969). Thus a high value of m indicates that searching efficiency declines markedly with consumer density whilst a low value indicates that interference is unimportant.

Interference can be considered as being over food or mates. If over food the values of m are usually small and less than one. If a number of females arrive at a site with the intention of mating, then the probability of a given male mating declines with the number of males present. If the time involved in mating is a small proportion of the total, and thus handling time is negligible, the mating rate (W) can then be expressed as:

$$W = N_f \, N_m^{-m} \tag{7.2}$$

where N_f is the number of females, N_m is the number of males and m is a measure of the strength of interference. Lek mating systems can be considered as a form of continuous input (see Section 3.7) in which the males compete to mate with each arriving female (Parker 1978). For such continuous input studies the value of m should equal 1 (Sutherland and Parker 1992) (see Chapter 3).

On leks there may not, however, be a simple relationship between the number of males and the average mating success. This is a consequence of disruption, in which males attack copulating pairs. Trail (1985) showed that disruption of mating was frequent on leks, and occurred in various species at levels ranging between 2% and 68% disruption of all attempted matings. Disruption is likely to be greater when males are densely clustered (Foster 1983). For example, Kafue lechwe *Kobus leche kafuensis* were disrupted more when attempting to mate in large groups than when in small groups (Nedft 1992). It is possible that disruption sometimes causes females to leave large leks without mating in which case the value of m will exceed 1.

Other factors may result in m exceeding 1. There are records of female mallards *Anas platyrhynchos* drowning due to harrassment by groups of males. Similarly in dungflies *Scatophaga stercoraria* male–male competition can result in females being drowned in the dung (Hammer 1941).

7.4 Individual differences

Many studies have shown that there are phenotypic and behavioural differences between males within a species and numerous studies of leks have emphasised the fact that certain males within the lek obtain a disproportionate fraction of the matings (Davies 1978; Mackenzie *et al.* 1995). In lekking species, females fight less often than males and may not even compete at all; hence it is less likely that individual differences among females are as important.

These individual differences can be incorporated into eqn 7.2 by assuming individuals will differ in the amount of interference experienced. Thus the mating success of poor competitors will decline more with density than will the success of good competitors. Following Sutherland and Parker (1985) and Parker and Sutherland (1986), mating success (W_i) for an individual of phenotype i is

$$W_i = N_f \, N_m^{-m \, R_i} \tag{7.3}$$

where R_i is the relative competitive ability expressed as the average competitive ability of all individuals in that site divided by the competitive ability of the individual under study. Thus a poor competitor will suffer particularly severe interference when surrounded by strong competitors yet experience average interference for the population when surrounded by equally poor competitors. In studies of red-backed voles *Clethrionomys rufocanus bedfordiae* the variance in male mating success is related to variance in the size of the voles (Kawata 1988).

7.5 Leks: theory

Leks are thought to occur when it is not economic for males to defend females or the resources the females require (Bradbury 1977; Emlen and Oring 1977). Resource defence is uneconomic where the females move over a wide area (Davies 1991). In a number of species (Uganda kob *Kobus kob thomasi*, topi *Damaliscus lunatus*, and fallow deer *Dama dama*) males defend either resource-based territories or harems when at low densities but lek when at higher densities, presumably because defending large territories is uneconomic with the high cost of excluding others (Clutton-Brock *et al.* 1988)

In this section I will explore whether the approach of the ideal free distribution described in the previous section can also be used to consider the distribution of males between leks. Vehrencamp and Bradbury (1984) stress the importance of game theory for furthering our understanding of leks and such models have been produced by Parker (1978), Bradbury *et al.* (1986), and Beehler and Foster (1988).

There are many other explanations for lekking such as the hotspot model, in which there are random overlapping home ranges of females and males displaying in the areas where the number of overlapping females home ranges is greatest (Bradbury *et al.* 1986). The prediction of such models is that leks should occur at a distance apart approximately equal to the diameter of the female home range. Wiley's (1991) review shows that in many species the female home range is often considerably greater than the distance between leks (Gibson and Bradbury 1986; Rhijn 1983; Svedarsky 1988; Wegge and Rolstad 1986). Furthermore, about a third of female sage grouse *Centrocercus urophasianus* do not visit the nearest lek (Bradbury *et al.* 1989).

Many studies have shown that the central male in leks obtains the most matings (Wiley 1991). Beehler and Foster (1988) reinterpreted this as suggesting that leks form around certain high-quality males, which they called 'hotshots'. Höglund and Robertson (1992) showed that temporarily removing a male great snipe *Gallinago media* from a territory on the periphery of the lek simply results in the vacated territory being occupied by another, while if a central male is removed the vacated territory is not occupied and the entire cluster of males disperses. This is compatible with the idea of males benefiting from being adjacent to an attractive male.

Leks often persist in almost exactly the same spot for decades or longer (Wiley 1991) which does not fit well with the hotshot idea of females collecting around particular high-quality individuals. Likely explanations for this persistance are either that females remember lek locations and return in subsequent years or that suitable lekking locations are restricted.

Another suggested explanation for leks is that females may show a preference for mating within aggregations of males. This may be because it makes mate choice easier, because such sites are safer from predators or because mates can be chosen more rapidly (Alexander 1975).

The black hole model (Stillman *et al.* 1992) assumes that females leaving a territory tend to move to another nearby territory. The model shows that, as a result of females moving to the nearby territories, isolated territories are rarely visited and thus males gain the highest number of mates from defending territories close to others. Thus males should not occupy isolated territories. With a benefit of being close to others the theoretical expectation is that isolated males will move to the centre and thus the distribution of males will coalese—rather like black holes—and result in leks.

Female Kafue lechwe outside territories experienced high levels of harassment from non-territorial males (Clutton-Brock *et al.* 1993) and such harassment may be important in many lekking mammals. Harassment could be a major reason why female mammals in oestrous move to a nearby territory when they leave another.

The main objective of this chapter is to explain the obvious fact that the displaying males occupy a minute fraction of the apparently suitable habitat available. For example, estimating the total area of all leks relative to the size of the available habitat apparently suitable for lekking gives the following percentages: great snipe 0.03%, black grouse *Tetrao tetrix* 0.12%, and ruff *Philomachus pugnax* 0.04% (J. Höglund, unpublished data).

One approach is to assume that the distribution of females is determined by the position of feeding sites, nesting sites, areas safe from predators or other resources. The distribution of males, in relation to a fixed distribution of females, may then be considered as an ideal free distribution (Bradbury *et al.* 1986; Parker 1978) with the number of males in a site, N_m, being related to the number of females in the site, N_f, by

$$N_m = c \, N_f^{1/m} \tag{7.4}$$

where c is a normalizing constant (Sutherland 1983). What are the expected consequences for the distribution of the sexes? If m is low, as when competing for food, then consumers will aggregate in the patches of highest prey density. If m is near 1, such as when males are competing for females, the expectation is that the males will be dispersed in direct proportion to the densities of females. As described earlier, in the absence of disruption, m should equal 1 and, from the above equation, we expect input matching with

$$N_m = c\, N_f \qquad (7.5)$$

(Sutherland and Parker 1985, 1992).

Thus the prediction of this approach is that the number of males in each site will be directly proportional to the number of females. The observed highly aggregated distribution of males thus requires an equally aggregated distribution of females. This may be the case for the aggregations of male insects linked to the aggregations of females which are themselves dependent upon limited resources such as oviposition sites, sunspots, female emergence sites, nest sites or adult feeding sites (Parker 1978). Alexander (1975) refers to these as resource-based leks. It could, however, be argued that this term is an oxymoron; if the females' distribution is determined by the resources then this behaviour cannot be included within the usual definition of leks.

It is thus easy to explain aggregations of invertebrates in which the resource is strictly limited and the males cluster around these patches. However, for vertebrates the females are widely distributed and it is clear that this is a insufficient explanation for leks.

In some cases females may be attracted to safe sites; Gosling and Petrie (1990) suggest that female topi are less likely to be captured by predators when on a lek as these have good visibility due to grazing. The mean mating success of Uganda kob increases with the distance of thickets from the lek but removing thickets had no effect on the distribution of males or females or mean mating success (Deutsch and Weeks 1992). A lion *Panthera leo* kill on one lek resulted in a decline in mating success of males on nearby territories and a rise in success of distant ones as a result of the movement of females (Deutsch and Nefdt 1992). Black grouse leks in Finland are largely restricted to bogs and ice-covered lakes. The distribution of resources can clearly play some role in determining the distribution of leks. It does, however, seem unlikely that this is sufficient to account for the use of such a small fraction of the available habitat as decribed earlier.

From eqn 7.4 it is clear that an increased level of disruption, expressed as a value of m exceeding 1, will result in individual males tending to disperse between patches rather than accumulating in a few (Figure 7.2).

We can then consider how males differing in competitive ability are distributed between leks visited by different numbers of females. The gains to males of each phenotype are expressed by eqn 7.3. Thus the mating rate of each individual male depends upon the number of females visiting the patch, the number of males in that patch, and the male's competitive ability relative

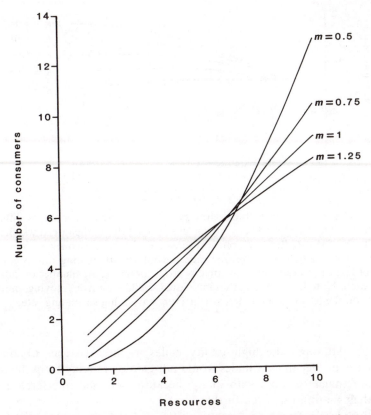

Fig. 7.2 The distribution of consumers in relation to the density of resources for different values of interference, *m* as expected from eqn 7.4. For individuals competing for food *m* is usually ≤ 1, for individuals competing for mates *m* = 1, and for individuals competing for mates with disruption *m* > 1.

to others in the patch. We can determine the evolutionary stable strategy by distributing the individuals in proportion to their mating success (see Section 2.4). At the evolutionarily stable strategy no individual of any phenotype gains by moving (Parker 1982). The distribution of males is likely to affect both the distribution of females and the position of other males (Alatalo *et al.* 1992; Beehler and Foster 1988).

With the evolutionarily stable strategy there are marked differences in the distribution of the different phenotypes with the better competitors accumulating in the sites with the most females (Fig. 7.3).

In the previous model without individual inequalities (Fig. 7.2), the number of males in each patch was directly proportional to the rate at which females arrived. As shown in Fig. 7.3, the incorporation of inequalities results in a more even spread of males between patches. As a result, the average mating success of males present on each patch increases with the number of females

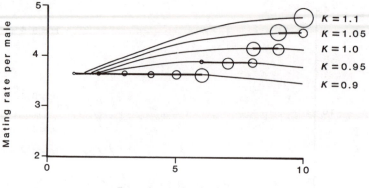

Fig. 7.3 The evolutionarily stable strategy of five different phenotypes differing in competitive ability distributed between ten leks which differ in the rate at which females arrive. The mating rate that each phenotype would obtain at each lek is shown by the plain line. Bold lines show where, for each phenotype, the mating rate is equal in a number of leks. The radius of each circle is proportional to the number of individuals in each patch. Note that, although no individual would gain from moving, individuals on the leks with more females arriving have a higher mating rate.

(Fig. 7.3). Although the high-quality males in these patches obtain more matings, the poorer-quality males would not gain by moving from the patches with few females to those with many, because increased interference would reduce their mating rate even further.

7.6 Leks: observations

In an explicit attempt to test the model of the ideal free distribution with unequal competitors just described (Fig. 7.3), Alatalo *et al.* (1992) organised a team to watch simultaneously nine leks of the black grouse in central Finland during the main period of matings (Fig. 7.4). They also collected data on the number of adults and juveniles on a further eight leks. Lek size varied between 2 and 20 males. On larger leks there were more female visits and more copulations both in total and per male than on smaller leks.

A study of ruffs on Gotland, Sweden, extended this idea to examine the link between habitat availability, female density, and male density (J. Höglund, F. Widemo, W.J. Sutherland, and H. Nordenfors, unpublished). The ruff leks were located on short grass on raised hummocks. Females selected areas of long grass to nest in, perhaps because the greater cover hides them from predators, and leks with a greater area of high vegetation nearby had more females visiting (Fig. 7.5).

The leks with more females visiting also had more males present. Thus male distribution was related to the extent of tall vegetation via the numbers of

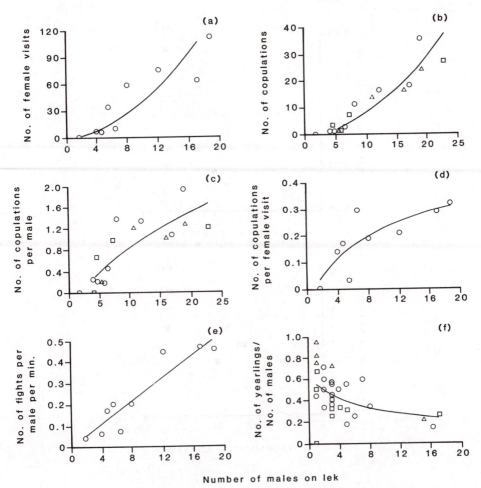

Fig. 7.4 The relationship between the numbers of males in each black grouse lek and various parameters (triangles = 1987, squares = 1988, circles = 1989; in graphs (a)–(e) number of males on the x axis includes yearlings). (a) The number of female visits ($r_s = 0.95$, $n = 9$, $P < 0.01$); (b) the number of copulations ($r_s = 0.98$, $n = 9$, $P < 0.01$); (c) the mean number of copulation per male (1989: $r_s = 0.87$, $n = 9$, $P < 0.01$; 1987–89: $r_s = 0.82$, $n = 17$, $P < 0.01$); (d) the proportion of female visits that resulted in copulation ($r_s = 0.85$, $n = 9$, $P < 0.01$); (e) the mean number of fights per male per minute ($r_s = 0.92$, $n = 9$, $P < 0.01$); and (f) the proportion of males on the lek that were yearlings plotted against the number of adult males (1989: $r_s = -0.4$, $n = 23$, $P < 0.05$; 1987–1989: $r_s = 0.47$, $n = 35$, $P < 0.01$). The regressions are drawn according to a best fit among linear and logarithmic models. (From Alatalo *et al.* 1992.)

females even though males never visited the nest or the breeding habitat. As with the black grouse, in accordance with the theory, the average mating success was higher on the leks with more males and thus in turn was also related to the extent of tall vegetation. Fighting rate also increased with the number of

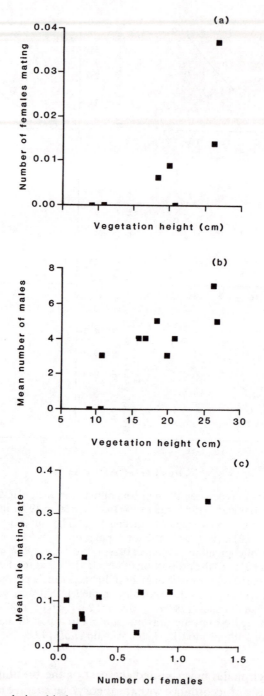

Fig. 7.5 The relationship between the amount of suitable nesting habitat surrounding ruff leks and (a) the number of females mating per lek (corrected for observation time) ($r_s = 0.79$, $n = 7$, $P < 0.05$), and (b) the number of males per lek ($r_s = 0.81$, $n = 81$, $P = 0.016$ (c), and the average mating success of the males in relation to the number of females mating ($r_s = 0.56$, $n = 13$, $P < 0.05$). (From J. Höglund, F. Widemo, W.J. Sutherland, and H. Nordenfors, unpublished.)

males. Thus, this study shows how the links between the resources, numbers of females, and number of males shown in Fig. 7.1 applies to leks.

The studies that have examined the link between lek size and mating success are reviewed in Table 7.1. No clear pattern emerges!

7.7 The origin of female preferences

There is thus evidence that both habitat preferences and female choice influence the size and location of leks. The framework proposed here leads to a possible explanation for female preferences.

The expected distribution of males of different quality between leks may provide a novel explanation of the evolution of leks. The expectation of the theory outlined in this chapter is that male quality is likely to be correlated with male density. This is supported by some of the field studies. This relationship between male quality and lek size could provide the explanation for the origin of female preferences of larger leks. Females visit the larger leks because that is where the highest-quality males will be.

For this preference to evolve, it initially requires an uneven female distribution resulting from processes such as variation in nesting or breeding habitat. Males are then distributed in leks in relation to female density but the leks visited by more females have more males, a greater proportion of higher-quality males and a higher average mating success per male. If male quality is heritable then there may be selection for females choosing the larger leks. We can then expect an accelerating female choice spiral (Parker 1978) in which both males and females benefit from attending larger leks and thus both sexes will theoretically collect in an ever restricted range of patches. The theoretical expectation of all individuals using a single lek will be counterbalanced by factors such as a limit on the distance females will move and limitations on the ability to find distant leks.

The rate at which females initially visited each site could be due to the density of females in the surrounding area or passive attraction (Parker 1982) in which they are just more likely to see or hear larger leks. However, in the study of black grouse (Alatalo *et al.* 1992) not only did more females visit larger leks but a higher proportion of these copulated (Fig. 7.5) which suggests that active female preferences for larger leks contributed to the increased mating success on larger leks.

Studies of some other species have also recorded a preference for larger leks. In studies of sage grouse, the addition of loudspeakers through which sage grouse calls were played attracted females (Gibson 1989). Similarly, in an experiment in the Netherlands, female black grouse were more likely to land next to a group of six models of male grouse than a group of three (Kruijt *et al.* 1972). Free-flying ruffs given a choice of groups of males in cages were more likely to join the cage with the higher density of males (Lank and Smith 1992). The number of females per male also increased.

Table 7.1. Studies that relate the mean mating rate or the number of visiting females to the number of males on a lek; this table also includes studies of invertebrates

Species	Observation (O) or experiment (E)	Relationship	Fit with theory	Author
Black grouse *Tetrao tetrix*	O	Higher mating success, a larger proportion of adults to juveniles, and more fights in larger leks	Yes	Alatalo *et al.* (1992)
Uganda kob *Kobus kob thomasi*	O	Ejaculation rate per male increased with lek size	Yes	Balmford *et al.* (1992)
Sage grouse *Centrocercus urophasianus*	O	Female visiting rate does not increase with lek size	No	Bradbury *et al.* (1989)
Uganda kob *Kobus kob thomasi*	O	No relationship between lek size and number of females per male	No	Deutsch (in press)
Greater prairie-chicken *Tympanuchus cupido*	O	Mating success does not increase with lek size	No	Hammerstrom and Hammerstrom (1955)
Stink bug *Megacopta punctatissimum*	O	Females court more in aggregations	Yes	Hibino (1986)
Ruff *Philomachus pugnax*	O	Higher mating success on larger leks, more fights on larger leks	Yes	Höglund *et al.* (1993) J. Höglund, F. Widemo, W.J. Sutherland, and H. Nordenfors (unpublished)

Table 7.1. Continued

Species	Observation (O) or experiment (E)	Relationship	Fit with theory	Author
Sage grouse *Centrocercus urophasianus*	E	Playing grouse calls increased the mating success of the males	Yes	Gibson (1989)
Black grouse *Tetrao tetrix*	E	Females more likely to join group of six models than three	Yes	Kruijt (1972)
Ruff *Philomachus pugnax*	E	Wild females prefer to join large group of captive males than small	Yes	Lank and Smith (1992)
Golden-headed manakin *Manucus manucus*	O	More visits by females and more copulations per male in larger clusters	Yes	Lill (1976)
Midge *Chironomus plumosus*	O	Mating rate of males highest in smaller swarms	No	Neems et al. (1992)
Drosophila conformis	O	Mating success does not increase with lek size	No	Shelly (1990)
Emphid fly *Emphis borealis*	O	Male mating success increases with size of male aggregation	Yes	Svensson and Pettersson (1994)

I thus suggest that the restricted distribution of females and a limited number of safe sites in which to mate results in an aggregated distribution of males. Small aggregations will contain poorer-quality males than larger agregations and this, in turn, will result in females selecting the larger aggregations within their home range. Such a mechanism could result in leks. The resulting size of the leks and their distribution will then depend upon the distribution of resources and lekking sites, the degree to which females select leks, and the distance they will move to visit a lek.

It is likely that a range of explanations are necessary to explain lekking and different explanations may apply at different scales. Hotspot models may explain the general distribution of leks, hotshots may explain the position of individuals within leks, while female preference and black hole models may explain why leks occur. Black hole models may be more relevant for mammals in which harassment is more important and female choice may be more important in birds.

7.8 Summary

The ideal free distribution with unequal competitors provides a reasonable description of lekking males. It may also help account for the distribution of females and thus leks. One prediction is that good competitors should tend to occur in the larger leks. This provides a possible explanation for female attraction to larger leks as a higher proportion of the males there will be of high quality.

Ruffs and black grouse were studied with the intention of testing the ideal free distribution with unequal competitors. For the ruffs, the number of females in each lek depended upon the height of the surrounding vegetation, where they nested. This, in turn influenced the number of males. For both species the larger leks had more females visiting and a higher mating rate per male.

8

Population regulation

8.1 Introduction

This chapter will show how population size can be understood in relation to aspects of behaviour such as interference, depletion, and territoriality. Population size results from the interaction of density-dependent and density-independent mortality and breeding output. In this chapter I will start by considering the links between foraging behaviour and the nature of density-dependent mortality and then between territorial behaviour and density-dependent breeding output, and then show how both of these can be combined to give the equilibrium population size. It is then possible to explore the relationship between behavioural measures, such as the degree of interference or territorial behaviour, and population size.

8.2 Density-dependent mortality

Earlier chapters describe how the competitive processes of interference and depletion can result in an individual's food intake declining with the density of consumers. Field evidence to show that both interference and depletion are important was outlined and a theoretical framework was developed for describing the consequences of these processes for intake and distribution.

When the population size of consumers is high, the intake of individuals will be reduced by two processes. First, the higher densities within patches will result in increased interference and depletion, and, second, a higher proportion will feed in the poorer patches. If there are individual differences in competitive ability then the poorer competitors may be expected to have a particularly low intake as they will both occur in poorer patches than the rest of the population and experience more interference when with better competitors.

It seems realistic to assume that individuals need a threshold intake to stay alive. Thus the reduced intake expected with high densities of consumers will result in more individuals falling below this level and starving. The consequences of this can be seen by returning to the model in Chapter 3 that shows how intake rate varies in relation to the level of interference, depletion, variation in competitive ability, and the number of consumers. This model can be extended to include the threshold intake for survival (Sutherland and Dolman 1994). The relationship between mortality rate and the number of consumers for a set of parameter values is shown in Fig. 8.1.

The nature of density-dependent mortality depends upon the values of interference, variation in competitive ability, and depletion rate (Fig. 8.1). Not surprisingly, a high level of interference or depletion results in starvation at a lower consumer density. If there are large differences between individuals in competitive ability then mortality starts to act at lower densities of consumers as the poorer competitors die first.

If there is depletion but no interference, then this results in a threshold consumer density below which there are sufficient resources and everyone survives and above which depletion reduces the resources below the threshold level and everyone starves; Nicholson (1954) referred to this concept as scramble competition. Thus, theoretically, the addition of a single individual to the population can result in total starvation. Once the resources have been depleted then the death of some individuals will have little effect on those remaining, provided there is no productivity.

There are a number of reasons why this absolute all-or-nothing starvation does not usually occur in practice. Firstly, individuals will differ in their ability to tolerate low intake due to, say, differences in condition or parasite load. For example, in the desert toad *Scaphiopus couchii* monogean parasites result in a drain of around 7% of the annual lipid requirement (Tocque 1993) and parasitized individuals weigh less. Thus the threshold intake necessary for survival differs between individuals. There is good evidence that individuals differ in their likelihood of dying. In most studies mortality is considerably greater for juveniles (Catterall *et al.* 1989) and subdominant individuals (Arcese and Smith 1985; Baker *et al.* 1981; Fretwell 1969; Glase 1973; Kikkawa 1980; Smith 1976). However, this process is not inevitable. Sullivan (1989) studied yellow-eyed juncos *Junco phaeonotus* and found that, despite the fact that starvation was the major mortality factor and that juncos have dominance hierarchies based on age and sex, there was no relationship between mortality and neither age nor sex. The same seems to be true for dark-eyed juncos *Junco hyemalis* (Ketterson and Nolan 1983).

Secondly, if some productivity of the food resource is occurring at the same time as depletion proceeds then, as some individuals starve, the subsequent productivity is shared between fewer individuals. Finally, intake may be influenced by interference and thus as some individuals starve the others will benefit from the reduced interference.

It is usually difficult to determine the strength of density-dependent mortality. Annual mortality rates are difficult enough to estimate and density-dependent mortality requires data from many years with varying population size. There are also considerable problems with analysing such data. For example the standard statistical techniques may often fail to detect density

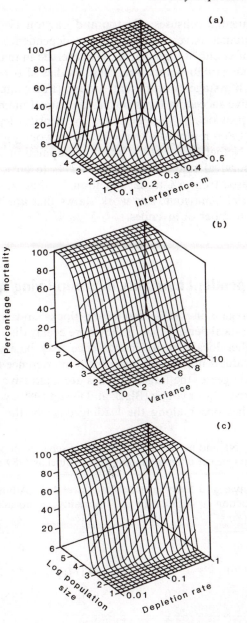

Fig. 8.1 The relationship between mortality and population size for various combinations of (a) interference—all individuals are of equal competitive ability, (b) variance in competitive ability—interference is constant $m = 0.2$, and (c) depletion—over 500 equal time intervals. For all simulations, mortality acts before the depletion is calculated. Unless stated interference is constant $m = 0.2$, variance = 4. Competitive ability is normally distributed over 21 phenotypes spanning ± 3 standard deviations around a mean of 10. Simulations consider 1000 consumers, feeding for 10 h per day for 150 days, in a site of 5 patches each of 20 hectares whose initial resource densities ranges from one to five items per square metre. Handling time = 20 s, quest constant $Q = 0.001$ m^2 s^{-1}. Threshold intake for survival = 0.001 items s^{-1}. (From Sutherland and Dolman 1994.)

dependence in sequential censuses (Gaston and Lawton 1987). Despite these problems, many studies have detected density dependence (see Table 8.1). Reduced survival of young at high densities is observed in many of the studies. However, as Sinclair (1989) points out, these studies also reflect the ease of studying different life stages in different groups. For example it is easy to measure reproductive success in birds but hard in small mammals.

Manipulations provide the best evidence for density dependence. As an example, Kluiver (1966) removed about 60% of the young great tits from the island of Vlieland, Netherlands, yet the breeding population size stayed stable, the adult annual survival rate increased from 27% to 56% and the juvenile survival rate increased from 6% to 22%. Although adults normally dominate juveniles in territorial behaviour, this work shows that adult survival is also determined by the number of juveniles.

8.3 Density-dependent mortality in group-living species

Many species aggregate in herds, shoals, or flocks and hence occur at a considerably higher local density than if they were evenly distributed over their range. At this higher local density interference may be important. Within groups, some individuals may suffer more from interference than others. For example, in barnacle geese, the larger families are aggressive towards smaller families and to those without young (Black and Owen 1989), which presumably explains why families occur along the leading edge of the flock where the

Table 8.1. Number (N), and percentage, of studies of separate populations demonstrating density dependence at different stages. (From Sinclair 1989.)

Group	Fertility/egg production N (%)	Early juvenile mortality N (%)	Late juvenile mortality N (%)	Adult mortality N (%)	Total no. populations
Fish	2 (6)	33 (94)	0	0	35
Birds	5 (26)	6 (32)	14 (74)	4 (21)	19
Small mammals	0	0	12 (92)	1 (8)	13
Large marine mammals	34 (83)	10 (24)	0	1 (2)	41
Large terrestrial mammals	49 (68)	35 (49)	1 (1)	12 (17)	72

biomass of vegetation is higher and single birds or pairs without young feed further back in the flock and have to feed on a sward depleted by others. By contrast, in feral pigeons *Columba livia* heavier individuals occupy the centre of the flock, are less vigilant, and obtain food at a faster rate (Murton *et al.* 1972).

One major reason why individuals join groups is the reduced risk of predation (Bertram 1978; Goss-Custard 1970; Götmark *et al.* 1986; Hamilton 1971; Jarman 1974) although there may be other advantages such as a greater ability to detect or exploit food or decreased time spent vigilant or that there are advantages to juveniles in delaying dispersal from the breeding group for a range of reasons (Alexander 1974; Emlen 1991; Pulliam and Caraco 1984; Ranta *et al.* 1993). This leads to the possibility that density-dependent mortality may then act through a complex interaction of depletion, interference, and predation. Consider the simple case of an increase in population size resulting in more groups but not larger groups. The greater number of individuals will result in greater depletion. Within groups, this food shortage will be compounded by localized depletion and interference further reducing intake. Those individuals which suffer the greatest interference may then obtain insufficient food within groups to stay alive. By leaving the group their intake may increase as they are no longer prone to interference and localized depletion, but they are then, however, more susceptible to predation.

The alternative possibility is that at larger populations the number of groups stays similar but they are larger. A similar argument applies, some individuals will experience greater depletion or interference and thus may leave and be more susceptible to predation. With either process, at high population densities more individuals may leave flocks and be eaten. Thus starvation may be rare, yet there may be density-dependent mortality linking the population size to the food supply, but this may be mediated through predation rather than starvation.

8.4 Density-dependent breeding output

Amongst vertebrates, density dependence regularly acts in the breeding season—reviewed by Sinclair (1989) and see Table 8.1. Sinclair's review shows that density dependence usually acts either through predation or as a decline in fecundity or egg production, largely due to limitations in the availability of space or food.

Territorial behaviour can then result in density-dependent loss in breeding output as described in Chapter 5. At high consumer densities, territories may be smaller so that fewer young can be raised, a higher proportion of territories may be established in poorer quality patches, and a higher proportion of individuals may forego breeding.

The relationships between mean breeding output and population size can be described by the model of Maynard Smith and Slatkin (1973).

$$Y = N \{1 + (aN)^b\}^{-1} \tag{8.1}$$

where Y is the number of individuals which breed; N is the number of individuals; a is a scaling constant which determines, for a given value of b, the density at which proportionate mortality reaches a constant value; and b is the strength of the relationship between density and the loss of breeding output. If the value of b is 1, as shown by a number of studies (Galbraith 1988; Newton 1986), then there is perfectly compensating density dependence; the total number of young produced by the population is the same over a wide density range. If b exceeds one, as shown by studies of elephants *Loxodonta africana* (Dobson and Poole, in press) then it is overcompensating—the larger the population the fewer young are produced. Other studies of a range of species have shown that total breeding output is proportional to density, i.e. there is no density dependence and thus b has a value near to zero. However, for most species the value of b seems to lie between 0 and 1 (Sinclair 1989).

Territorial models may be used to predict the nature of density-dependent breeding output, as shown in Chapter 5. In that chapter, I showed that at higher densities of consumers we should expect more individuals to act as floaters rather than to breed, individuals to occupy smaller territories, and the occupancy of poorer patches. Field studies also show that these phenomena are more important at high densities (Chapter 5). The value of b will thus depend upon aspects of the behaviour. For example, if territories are occupied in order of quality then the extent of density-dependent breeding output depends upon the extent of variation in territory quality (Fig. 5.3). Also, the relationship between population size and the number of individuals queueing rather than breeding depends upon the survival rate and variation in territory quality (Fig. 5.5).

8.5 Equilibrium population size

The equilibrium population size is reached when mean breeding productivity is balanced by mean survival (see Fig. 1.5). This can be determined by combining the density-dependent mortality shown in Fig. 8.1 with a given level of density-dependent breeding output, as in Fig. 8.2. This model considers a population that has a single wintering area and a single breeding area. The next chapter extends it to consider a population which uses a range of sites.

Figure 8.1 shows that the nature of density-dependent mortality varies with the extent of interference, depletion, and variation in competitive ability, it is thus possible to relate these processes to equilibrium population size. Figure 8.2 shows how population size relates to the different degrees of density-dependent breeding output. With very high values of density-dependent breeding output, b, the processes that affect density-dependent mortality have little effect on population size. However, under the more usual conditions of b being less than one, population size is related to the behavioural variables. At higher

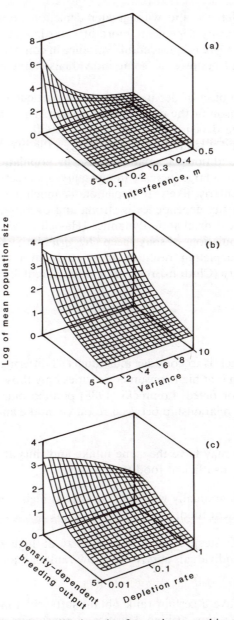

Fig. 8.2 The equilibrium population size for various combinations of both density dependence in the breeding grounds and (a) interference, (b) variance in competitive ability, and (c) depletion. Winter site and foraging parameters and threshold intake rate are as in Fig. 8.1. The mean post-breeding population is calculated over 40 generations after allowing 160 generations for the simulation to stabilize. The population breeds in a site of 25 km², $a = 0.25$; an equal sex ratio is assumed; fecundity per female = 1.2 per annum in the absence of density dependence. Density-independent mortality (0.15 per annum) acts on the post breeding population before the winter. (From Sutherland and Dolman 1994.)

strengths of interference and with greater depletion rates the population is smaller, due to density-dependent mortality starting to operate at a lower population size. If there is considerable variance in competitive ability then the population size will be lower as some individuals starve at lower population sizes.

One common subject of debate is whether a particular species is regulated in the breeding season or the non-breeding season. This approach shows how the population size depends upon processes in both the breeding and non-breeding season and it is the interaction between the two that is important.

Under some conditions a small increase in population density may be expected to result in a large increase in mortality. As described earlier (Section 8.2) this is particularly likely if depletion is much more important than interference. If density dependence is strong and overcompensating then this can result in cyclical or chaotic dynamics (Hassell *et al.* 1976). The cyclical fluctuations of populations of soay sheep *Ovis aries* on St Kilda, Scotland, can be explained by depletion resulting in strongly overcompensating density dependent mortality (Clutton-Brock *et al.* 1991; Grenfell *et al.* 1992).

8.6 Rank

A different approach is to consider individuals as differing in rank (Łomnicki 1978, 1988) such that at high consumer densities only those individuals of high rank can survive or breed. Łomnicki (1988) pointed out that there are four ways in which the relationship between resource intake and rank may change with density:

(1) all individuals may have the same intake and thus at high densities they will all obtain insufficient food to stay alive;

(2) intake declines gradually with rank so that at high consumer densities all obtain insufficient food;

(3) intake declines steeply with rank so that at high consumer densities high-ranking individuals survive and low-ranking individuals obtain very little food;

(4) individuals above a certain rank obtain sufficient food and those below this rank starve, thus at high densities of consumers a higher proportion starve.

These four divisions are really points on a continuum ranging from scramble competition to contest competition. They differ in the extent to which low-ranking individuals influence the intake of high-ranking individuals and this ranges from them having a considerable effect in (1) to having none in (4). As a result, (1) produces considerable resouce depletion at high-consumer densities which may result in overcompensating mortality, while (4) results in a large population and greater population stability.

The results of these four options described by Łommicki are similar to those of the ideal free distribution with all individuals equal [equivalent to (1)] ranging to considerable variation in competitive ability [equivalent to (3) or (4)]. Similarly the ideal despotic distribution will produce comparable results, with (1) being equivalent to no territoriality and (4) being equivalent to inflexible territory sizes. Although the results are similar it is sensible that each field system is modelled by the theory that describes the behaviour most closely.

An example of how rank can be related to population size, is the approach I used with Heribert Hofer in modelling the population of hyenas. As described in Chapter 3, rank has considerable consequences for intake of hyenas at kills. In many carnivores, low-ranking females are unlikely to breed (Macdonald 1979). This relationship between rank and reproductive success has been quantified for hyenas (H. Hofer and M.L. East, unpublished). It can be considered as a form of density-dependent fecundity as at high population density, a greater proportion fail to breed successfully. We determined the equilibrium population size to result from various different density-independent mortality rates in combination with the density-dependent breeding output. We could then consider the consequences of changes in mortality. For example, in recent years more hyenas have died in poachers' snares and it was possible to show that time necessary for recovery of the population from this mortality was considerably longer than previously expected.

8.7 Carrying capacity

Carrying capacity is a term with many contradictory meanings such that some have suggested the term should be abandoned (Dhondt 1988). Goss-Custard (1985), Goss-Custard and Durell (1990), and Sutherland and Goss-Custard (1991) suggest that, for birds wintering on estuaries, it is useful to use the term carrying capacity for those cases where the addition of a further individual will result in the death or emigration of another. Using this definition it becomes possible to determine the population a location can sustain (see Chapter 10 for examples). From the depletion model in Chapter 3 it is straightforward to determine the carrying capacity and Fig. 3.3 shows how to determine the number of individuals that can be sustained for the duration of the winter.

There can, however, be problems with this definition of carrying capacity, as pointed out to me by Paul Dolman. The problem arises when there is a gradual increase in mortality with population size. Figure 8.3. shows an example of how mortality rate may increase with the population size within a site. Assume that individuals in this site come from a range of breeding sites so there is no relationship between the population in this site and the breeding output. The population size at which the addition of a further individual results in the death or exclusion of another is the point at which the slope equals one,

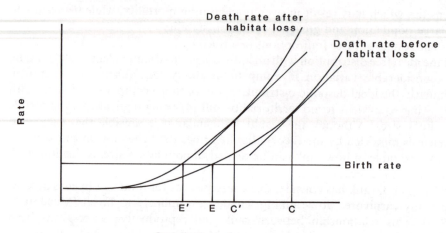

Fig. 8.3 The relationships between birth and death rates and population density within a site. The carrying capacity is sometimes defined as the population at which the addition of a single individual results in the loss or emigration of another. This can be considered as equivalent to a slope of one in this figure. The carrying capacity is thus the level C. After habitat loss has occurred the density dependent mortality rate is shifted to the left and the carrying capacity is now C′. The equilibrium population size is the point at which the birth rate equals the death rate and the value before (E) and after (E′) habitat loss is shown. Note that both values of equilibrium population size are smaller than both values of carrying capacity.

as one individual is then lost for each one added. This would then be defined as the carrying capacity and this population size is shown by C in the figure. From this figure we can also examine the equilibrium population size. This is achieved by including the birth rate of the population and the equilibrium population, E, is the size at which mortality rate and birth rate are equal. This example shows that the equilibrium population size may be well below the carrying capacity.

Such a discrepancy between equilibrium population size and carrying capacity may have real consequences for conservation. As the equilibrium population is well below the estimated value of carrying capacity, this may well be used to argue that parts of the site may be lost without affecting the population. However if part of the site is lost the relationship between mortality rate and population size can be expected to change, as shown in Fig. 8.3, and will result in a new equilibrium population. Thus although in this example the population is well below the carrying capacity, a loss of part of the site still results in a reduced population.

What can then be said about the number of individuals a site will contain? The approach I suggest is unfortunately rather complex and it requires the question to be tackled in two stages. First it is necessary to consider the total population size resulting from the interaction of all the density-dependent and density-independent mortality and breeding output in all breeding sites and all non-breeding sites (the next chapter provides an approach for migratory species). The second stage is to consider how the individuals will be distributed between the sites. The number of individuals in a given site then depends upon the interaction of these two stages.

Determining the consequences of environmental change such as the destruction of part of a site requires the recalculation of these two stages. The loss may lower the equilibrium population and also result in a lower proportion using the remainder of the site. The combination of these determines the change in numbers. In reality, species may also take some time to respond and Sections 9.4 and 9.5 consider this issue.

8.8 Variation in food supply and weather

The aim of this book is to show how behavioural processes such as depletion, interference, or territorial behaviour have consequences for the population size. For simplicity the models described assume that the environment is constant, and under such conditions we may expect, for example, a smooth relationship between mortality and the size of the consumer population. Furthermore, such relationships will be identical each year.

The real world is, of course, very different and the environment will vary over time. The most obvious source of variation is fluctuations in the weather which may have consequences for the resource population, the ability of consumers to find resources, and the food intake necessary for individuals to survive. Other possible sources of variation include natural or human induced changes in the resource population, changes in parasite loads in either the consumers or resources, and changes in the degree of interspecific competition or predation.

Figure 8.4 illustrates the possible consequences of variation in resources or weather. It shows a likely range of density-dependent mortality responses expected in typical years and an example of a poor year in which the resources are few or the weather is severe. In periods of resource shortage or severe weather density dependence may act strongly, but it may be minimal in periods of abundant food or during typical weather. Much of the population regulation may take place on infrequent occasions, and during what may be considered as atypical events. The population may take many years to recover and so such events may affect the mean population size.

The abundance of algae suitable for marine iguanas *Amblyrhynchus cristatus* in the Galapagos Islands declined dramatically following the 1982–83 El Niño–southern oscillation event. Survival of the iguana was then greatly reduced and there was evidence for density-dependent mortality acting particularly upon juveniles (Laurie and Brown 1990).

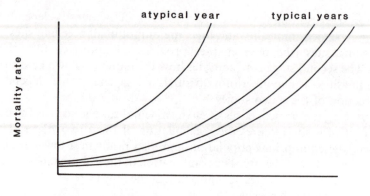

Fig. 8.4 How the relationship between mortality and consumer density may vary between years. A range of typical years (for example with slight differences in weather) and an atypical year (with severe weather) are shown.

The food supply of wading birds in the Wadden Sea in the Netherlands varies considerably among years (Beukema *et al.* 1993) such that resources may be plentiful in some years and much scarcer in others. This variation in food supply is caused by both annual variation in winter mortality and recruitment. The populations of many prey species fluctuate in synchrony and over large areas as a result of processes such as low over-winter survival during severe winters and high recruitment after severe weather, as either competition with adults is reduced or the predator numbers is reduced. This thus restricts the possibility of consumers switching prey species or moving elsewhere (Beukema *et al.* 1993). In some of these prey species the number of individuals settling from the planktonic stage (the spatfall) is very irregular. Thus in the Wadden Sea over a ten-year period there was reasonable spatfall of *Macoma balthica* in six of the years but *Mya arenaria* had a significant spatfall in only one year. There was no significant spatfall of *Scrobicularia plana* during the ten years of the study but a large recruitment had occurred the year before. As only the younger bivalves are within the depth and size range available to wading birds there is substantial annual variation in the density of available food (Zwarts *et al.* 1992).

In many consumer species, winter survival appears to be high and hence it may initially seem unlikely that density-dependent mortality can be an important process. However, mortality may be concentrated into very short periods of severe weather or food shortages. Black-capped chickadees *Parus atricapillus* provided with supplementary food during the non-breeding season had a higher average body mass and a higher overwinter survival rate. The increase in survival rate was largely due to a greater ability to survive periods of severe weather (Brittingham and Temple 1988).

After a cold spell in February 1991 reaching −12°C in one site, 2934 wader corpses were found on the Wash estuary, England, (one of the most important estuaries for wintering waders in Europe) of which 1553 were redshank. From counts it was estimated that 68% of the redshank population had died (Clark *et al.* 1993). In the next winter there was a higher proportion of juvenile redshank there than in other parts of the country or on the Wash in other years. This suggests, as observed elsewhere (see Section 9.4), that juveniles select the best sites while adults tend to be conservative and return to the same site. Thus, with a high mortality of adults on the Wash there is reduced interference and depletion which may be why so many juveniles were able then to settle.

If severe weather mortality in an area results in many of the juveniles settling in that area this may have consequences for juvenile recruitment elsewhere and may result in a reduced population.

8.9 Long-term consequences

The models of depletion presented so far assume resources are replenished every year, as, for example, would be the case for birds feeding on berries. In reality, under many situations the depletion may have long-term consequences for the abundance of the resources. As an example, studies of the brent geese in North Norfolk, England, feeding on *Salicornia* show that the geese tend to aggregate in the patches with the highest densities of the plants. The population dynamics of *Salicornia* in North Norfolk are well understood and it is known that *Salicornia* is regulated by density-dependent fecundity (Watkinson and Davy 1985). The consequences of additional mortality from goose grazing can then be considered by incorporating both processes in a population model (J.M. Rowcliffe, A.R. Watkinson, and W.J. Sutherland, unpublished). As shown in Fig. 8.5, the equilibrium population of *Salicornia* is reduced by the goose grazing. Current levels of goose grazing result in a population size about 15% lower than it would be otherwise. As well as altering the average abundance, goose grazing may also influence the dynamics of the plant populations. Brent geese aggregate in the patches with higher plant densities which can result in over-compensating density-dependent mortality. Figure 8.5 shows that with low levels of grazing there is a single equilibrium population but with slight increases in the extent of goose grazing the *Salicornia* population is expected to show cycles of abundance between years as shown by the two population sizes. With further increases in goose grazing chaotic fluctuations are predicted as is shown by the large number of population sizes in Fig. 8.5.

There is clearly the potential for much more research linking the behaviour of consumers with the dynamics of the resources.

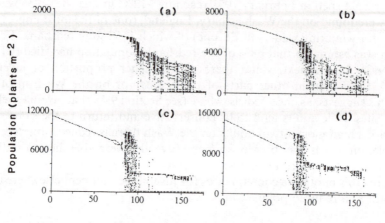

Fig. 8.5 The relationship between the population size of *Salicornia* and the extent of goose grazing. This shows the results for four levels of density-independent survival of the *Salicornia*: (a) 0.015, (b) 0.06, (c) 0.105, and (d) 0.15. Note that as the grazing pressure increases the population size shifts from a single stable point, to a pair of points (i.e. population cycles), to a large number of points (i.e. chaos). (From J.M. Rowcliffe, A.R. Watkinson, and W.J. Sutherland, unpublished.)

8.10 Summary

By assuming that predators starve if they fail to obtain a threshold intake, it is possible to determine the nature of density-dependent starvation. The nature of density-dependent birth rate may be dependent upon territorial behaviour. These may be combined to determine the relationship between the equilibrium population size and behavioural parameters. In some cases it is possible to say something useful about carrying capacities, but in most cases it is better to think of equilibrium populations. There may be considerable differences between years so that much of the density-dependent mortality may occur in occasional years with severe weather or low-resource populations.

9

Migration

9.1 Introduction

Migration, the return movement of individuals, occurs in many temperate and tropical birds, fish, turtles, and some mammals, such as whales, ungulates, and bats. The scale of migration may differ between long-distance migrations undertaken by many baleen whales, turtles, and seabirds, and localized movement such as roach *Rutilus rutilus* undergoing annual migrations within a lake to spawn in the same tributary (Vøllestad and L'Aee-Lund 1987).

The objective of this chapter is to show how it is possible to combine aspects of population ecology in both the wintering and breeding areas to determine evolutionarily stable migration routes. The aim is not to provide realistic models of precise migration routes, but to show the general manner in which migration may be understood. Such an approach assumes that individuals follow the best possible route and are not constrained by history or a lack of genetic mutations. In this Chapter I will thus examine the evidence for and against such constraints.

Evolutionarily stable strategies can only be produced if there is some negative feedback such as from intraspecific competition within sites. This assumption has been the basis of a number of models of migration (Alerstam and Enckell 1979; Alerstam and Högstedt 1982; Cox 1985; Fretwell 1980). Kaitala *et al.* (1993) give the most comprehensive analysis of this problem.

The simplest case of a migratory population is one that moves between a single wintering site and a single breeding site. For example, the barnacle goose *Branta leucopsis* may be divided into three well-delimited populations each with separate breeding areas, wintering areas, and migration routes; there is negligible interchange between populations. The population model of the previous chapter which assumes a single wintering site and a single breeding site applies directly to such cases.

However, in most species the populations will not be isolated. For example, many estuaries, such as the Wash in eastern England, are used by three different races of dunlin *Calidris alpina* each with a distinct breeding range; and many other species, such as fox sparrows *Passerella iliaca* consist of populations which differ in breeding areas but overlap in wintering sites (Mead 1983). To produce migration and population models of such species it is then necessary to consider the fact that different breeding populations will compete on the same wintering grounds.

Previous chapters have considered the factors determining the distribution of individuals between patches within a site. Much the same approach can be used to understand what determines the distribution between sites. The approach is to consider the choice of wintering sites and the choice of breeding sites, taking into account the cost of moving between them. This approach will produce the evolutionarily stable migration routes. It is thus just a question of changing the scale to shift from migration between local sites to movement between continents. As in the choice of patches within a site, resource abundance may not be the only factor influencing the choice of wintering sites. Other factors such as weather, predation risk, or historical constraints may also be very important (Alerstam 1990).

The essential elements of applying evolutionarily stable strategies to the choice of patches within sites is that individuals respond to both the resource distribution and the distribution of conspecifics. There is, also, evidence that both of these processes still apply when considering a much larger scale such as the choice of sites. Goss-Custard *et al.* (1977) showed that for a range of species there was usually a good correlation between the density of wading birds on different estuaries in south-east England and the densities of prey in each estuary. The numbers of grey plover wintering in Britain has been increasing and the increase has been greatest on those estuaries that previously held a lower density than expected (Moser 1988).

The ecological and behavioural processes that account for local scale movements may also apply on a larger scale. The seasonal pattern of habitat use of wintering brent geese between adjacent algae beds, salt-marsh, and agricultural land can be explained by the seasonal pattern of depletion, productivity, and natural mortality of the vegetation (see Chapter 3). Similar processes have been used to partly explain the regular migration of ungulates over considerably greater distances in East Africa (Fryxell and Sinclair 1988).

9.2 Migration costs

As Fretwell (1980) points out, it is remarkable how many species do not migrate given the fact that all environments are seasonal and it is likely that during some seasons the conditions would be better elsewhere. There must be disadvantages in moving. Migration seems obviously costly but these costs are hard to quantify. One obvious cost is the simple energy required to migrate. Some species of birds double their weight before migrating and the carrying of this extra fat is also thought to have costs, such as causing the bird to be less agile in avoiding predators (Witter and Cuthill 1993).

Migration may often involve moving across inhospitable habitats. Exhausted birds regularly fly into the sea or are caught by humans or other predators while resting on land. The large numbers of birds found on ships and gas platforms suggest that many more do drown. Alerstam (1990) suggests that millions of birds are displaced eastwards from North America each year and drown. Giant Aldabra tortoises *Geochelone gigantea* that migrate across relatively unshaded sites to the coast may die of heat stress (Swingland and Lessells 1979).

The costs may depend upon the movement of the medium in which they migrate. For bar-tailed godwits migrating from West Africa to the Netherlands in the spring, the average tailwind is 15 km h^{-1} but among different years this varied between a tailwind of 36 km h^{-1} to a headwind of 2 km h^{-1} (Piersma and van de Sant 1992). Fish similarly take advantage of tidal currents (Metcalfe *et al.* 1990) and the cost of migration can be calculated in terms of lost fecundity per kilometer moved.

The actual costs of migration in terms of mortality are very difficult to estimate. The best evidence comes from studying colour-marked barnacle geese in Svalbard before the autumn migratory flight and then in Scotland just after they arrived. Juveniles with a low body mass were less likely to survive migration. The best data is for 1986 in which winter arrived earlier than usual. In this year, 35% of the juveniles died during this period (Owen and Black 1989).

9.3 Evolutionarily stable migration strategies

Migration strategies cannot be considered in isolation. The approach adopted here is to consider the migration route of individuals taking into consideration the migration routes of other individuals both from the same breeding site and different breeding sites. In each wintering site there will be interference and depletion so that the survival probabilities will be density dependent. On the breeding grounds the breeding success will also be density dependent.

The approach adopted here considers a range of wintering sites and a range of breeding sites. Wintering sites are assumed to differ in size, but for simplicity the resource density is assumed to be the same in all sites. Breeding sites are assumed to differ in size but the nature of density-dependent breeding output is assumed to be the same in all sites. We assume that individuals migrate to a single site for the duration of the winter, within which their survival depends on the nature of depletion and interference. Of course, in reality resources are likely to vary between wintering sites and different breeding sites will differ in the relationship between breeding output and density due, for example, to the number of breeding sites or the resource abundance. It would be straightforward, using the model below, to vary any of these parameters between sites and determine the evolutionarily stable migration route.

If there is no migration cost, then the location of a site will not influence the extent that it is used and the depletion and interference models used in earlier chapters can be used to describe the distribution of individuals between patches across sites. This may be the case for sites located close together.

If migration costs are high then this may modify the choice of sites. Such costs can be incorporated either by assuming that each route has a set energy cost which requires additional food intake or a set mortality risk. In the model described here (Sutherland and Dolman 1994) the migration cost is incorporated by assuming that both the autumn migration, and the need to build up fat and protein reserves prior to spring migration, impose an additional requirement for food intake on the wintering site.

Such a model could also be developed to consider a range of routes and stopover sites between wintering and breeding sites. Even during the winter, movement may still occur between sites as shown in many species, such as the marsh warbler *Acrocephalus palustris* (Berthold 1988) and the yellow wagtails *Motacilla flava* (Wood 1979).

In this model, we assumed that the population is divided into phenotypes, which differ in competitive ability. As with the evolutionarily stable strategy described in earlier chapters, the solution is sought in which no individual of any phenotype in any population could gain by adopting a different migration route.

Figure 9.1 shows the expected migration pattern for a range of breeding and wintering sites. For each phenotype in each breeding population, the relative trade-offs between alternative winter sites are used to determine their distribution between winter sites in the following winter. If there is competition in the winter sites, the migration route adopted is a frequency-dependent process affected by the migration strategies of both other phenotypes from the same breeding population and populations from other breeding sites. The resulting pattern is an evolutionarily stable strategy in which no individual could gain by moving.

This model makes the reasonably simple assumption that the cost is proportional to distance. However, different migration routes may differ in cost. For example, the energetic costs of different routes for bar-tailed godwits depends on wind speed. In bar-tailed godwits, the spring migration from West Africa to the Wadden Sea in Europe in a single flight is only possible with a tail wind (Piersma and Jukema 1990). Thus the global pattern of winds may well influence migration routes.

Such a model can be of use in understanding the pattern of migration routes. As shown in subsequent chapters it can also be used to understand the consequences of habitat loss for populations. It is thus necessary to examine whether migration strategies are flexible or fixed.

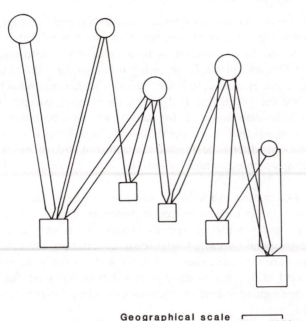

Geographical scale ⌐—————⌐
0   500km

Post—breeding numbers using ⌐—⌐
route scaled by arrow width 0 2000

Fig. 9.1 The migration routes with a range of breeding sites ○ and a range of wintering sites □. This shows a hypothetical geographical arangement of sites. The distances between the sites and the sizes of these sites are shown to scale. The line widths are scaled to show post-breeding numbers taking each route. For each winter site, in each year, available resources vary from one to five items per square metre in five equal-sized patches. For all breeding sites $a = 0.25$, $b = 0.2$, breeding output per female = 1.2 per annum in the absence of density dependence. Migration cost = 5×10^{-5} items per second per 1000 km. Foraging parameters and threshold intake for survival is as in Fig. 7.1. (From Sutherland and Dolman 1994).

9.4 Ability to change routes

The approach outlined here assumes that individuals move according to the profitability of different migration routes and there is the capacity to alter routes. Migration routes are likely to have regularly changed as a result of continental drift and, more recently, ice ages. Most of the migration patterns seen today clearly arose since the last ice age 10 000 years ago (Alerstam 1990). In this section I will consider whether we can assume that migratory populations have adopted the best route or whether they are constrained to adopt routes that were once sensible but no longer the best option.

Species differ markedly in the relative importance of genetic and cultural inheritance of migration routes. In some species with prolonged parental care, such as geese, the adults and young migrate together and the migration routes seem to be learnt; the young of pinioned geese do not migrate even if given the freedom to do so. Migration is also culturally determined in cranes (Erickson and Derrickson 1981). In many other species, the routes seem to have a genetic basis. A clear case is the cuckoo *Cuculus canorus* whose parents fly south before the young are fledged by their foster parents; their ability to migrate and the direction and distance flown must be innate. Terrill and Able (1988) suggest that it is useful to distinguish between the obligate phase of migration, which is performed regardless of environmental conditions and the facultative phase, which is induced by deteriorating environmental conditions.

The best documented example of change in route is the blackcap *Sylvia atricapilla*, which rarely used to be seen in Britain in winter, but since the 1960s has become a regular wintering species in some parts of the country (Lack 1986). Blackcaps spend the early winter feeding on natural food such as berries but, once these are depleted, they switch to feed on the wide range of food provided on bird tables in gardens. On this diet of birdseed, fat, bread, and other foods they can maintain or even increase their weight over the winter (Leach 1981).

Blackcaps are a common British breeding species so it was initially assumed that some of these breeding birds had simply switched to becoming resident. However, of the blackcaps recovered in Britain, almost a hundred were ringed on the continent during the breeding season and, moreover, none of those recovered in winter was from the British breeding population. It is thus clear that many continental birds have adopted a novel west or north-westerly migration route which results in them wintering 1000–1500 km further north than their traditional wintering area around the western Mediterranean. Despite thousands of German and Austrian blackcaps being ringed in the breeding season, prior to 1961 no west or north-westerly migrants were recorded, but they now account for 7–11% of the breeding population in parts of these countries (Berthold *et al.* 1992).

To study this switch in orientation, Berthold *et al.* (1992) captured some birds wintering in western England and returned them to Germany. The standard way of recording migratory orientation is to record their activity, known as migratory restlessness, in circular cages at night during the migratory period. The direction and duration of migratory restlessness is known to correlate well with actual migratory patterns. Not suprisingly, these British birds showed migratory restlessness in a direction towards Britain in the autumn while those from a population from south-west Germany showed a south-west preference in accordance with their migration to the western Mediterranean.

To distinguish whether this novel behaviour had a genetic or learnt basis, Berthold *et al.* (1992) allowed the British wintering birds to breed with each other and examined the migratory behaviour of the offspring. The migratory restlessness of the offspring was also directed towards Britain showing that the switch in orientation has a genetic basis. Furthermore, siblings showed more similar orientation than did unrelated young, which suggests a genetic component to the variance in orientation. Crossing birds that flew south-west (angle 227°) with those that flew to Britain (273°) resulted in offspring with an intermediate direction (253°) (Helbig *et al.* 1994). This showed that there was no evidence for a dominant allele resulting in the migration route to Britain. Furthermore, the orientation was the same regardless of whether it was the mother or father that adopted the route to Britain which excluded maternal effects.

For such a rapid change to take place, the selection pressure must have been strong. One possible reason for the switch to the new route is to take advantage of the increased habit of providing food on bird tables. Alternatively, in Britain the winter temperatures have increased in recent years, perhaps due to global warming, and it is possible that the switch is a response to this.

Work by Berthold (1993) and his colleagues has shown that both the direction and extent of migratory behaviour has a strong genetic component, and breeding experiments have showed that rapid changes in behaviour could theoretically take place. This switch to Britain shows that dramatic changes in migration are possible within decades.

There have been other examples of rapid changes in migratory behaviour. In the present century, serins *Serinus serinus* have spread from the Mediterranean to northern and central Europe and then switched from being sedentary to being migratory (Berthold 1993). An increasing proportion of great crested grebes *Podiceps cristatus* in the Netherlands (Adriaensen *et al.* 1993) and blackbirds in Central Europe (Heyder 1955) are now sedentary rather than migratory.

Such changes in routes require that mutations regularly arise that result in birds adopting new routes. Records of birds that have turned up remarkably off-course do occur, not infrequently. The writing of this book in East Anglia in the spring of 1993 was made more enjoyable (but slower) by the presence in East Anglia of a sociable plover *Vanellus gregarius*, which breeds in central Asia but winters in east Africa, and an oriental pratincole *Glareola maldivarum*, which does not usually occur nearer than India. Most of the vagrant birds that appear a long way from their usual route are juveniles. One possibility for this is that juveniles are inexperienced and more likely to get lost or blown off course. The other possibility is that these are mutants adopting a different migration direction (with the extreme vagrants also being helped by winds or boats) and the fact that few adults are seen suggests that these strategies are not successful.

Changes in route may also occur in species whose routes are learnt. Red-breasted geese *Branta ruficollis* have moved their wintering area from the Ukraine to Romania and Bulgaria following habitat changes (Sutherland and Crockford 1993). In the 1960s almost all the pink-footed geese *Anser brachyrhynchus* in Britain wintered in Lancashire and Scotland with only a few hundred in East Anglia, yet in recent years as many as 42 950 have wintered in East Anglia. Some barnacle geese *Branta leucopsis* have shifted to nest on the Baltic island of Gotland some 1 300 km south of their usual breeding range (Larsson *et al.* 1988). Within 200 years of being introduced to Britain, Canada geese *Branta canadensis* have set up a novel migration from central England to the Beauly Firth in north east Scotland where they moult, and then return (Walker 1970).

The shift of populations between wintering sites may be largely due to exploratory movement by juveniles. Many studies have shown that juveniles move more than do adults (Baker 1978). The manner in which such changes may take place is illustrated by the response of oystercatches to the dramatic spatfall of cockles on the Ribble estuary, England, in 1975; in the following winter the cockle population of the size range taken by oystercatchers had increased by an estimated 300-fold and the oystercatcher population that winter increased fourfold (Sutherland 1982*b*). The majority of these birds were juveniles or immatures. In subsequent years the populations of both cockles and oystercatchers declined and very few juveniles or immatures were present. The obvious interpretation is that juveniles explore and settle in areas with fewer consumers than expected for the level of resources while adults are more conservative and return to the sites used previously. Studies have shown that the distribution of birds between sites is related to the distribution of prey (e.g. Goss-Custard *et al.* 1977) and it seems likely that this may be largely due to the selective settling patterns of juveniles.

Other factors, however, may also influence migration routes. For example, in greater flamingos *Phoenicopterus ruber* that breed in a colony in France, there are consistent differences in the relative proportion of age classes between birds wintering in either Tunisia or Spain. Those that occurred in one country in the first or second winter tended to return to that country in further years (Green *et al.* 1978). The direction that is adopted in the first year seems to be related to the wind direction at the onset of migration in the bird's first autumn.

9.5 Constraints to changes in migration routes

The studies described in the previous section indicate that there can be a rapid response to change migration route when conditions alter. In this section I will describe evidence that some species seem to have failed to adopt migration routes that would seem better. This suggests there may be constraints in changing routes.

The routes adopted by some migrants may be best interpreted in terms of the migrants expanding their breeding range but maintaining their wintering range. This may result in apparently very inefficient routes. All wheatears *Oenanthe oenanthe* winter in Africa even though the breeding range has spread as far west as eastern Canada and as far to the east as through Asia to Alaska. In remaining faithful to their African wintering quarters the North-American-breeding birds may fly twice as far as they would if they migrated to South America. Similarly, all red-breasted flycatchers *Ficedula parva* winter in Asia although they now breed across Europe, and pectoral sandpipers *Calidris melanotos* breed well into northern Siberia but, along with the bulk of the population which nests in arctic America, these winter in southern South America. By contrast some other species with extensive breeding distributions have split their migration routes so that the population uses different wintering areas. For example, red-backed shrikes *Lanius collurio* in the west Palaearctic migrate to Africa whilst those in the east migrate to South-East Asia (Perrins and Birkhead 1983).

The entire European populations of lesser whitethroats *Sylvia curruco*, wood warblers *Phylloscopus sibilatrix,* and red-backed shrikes fly south-east around the eastern Mediterranean to reach Africa although for many individuals this is a longer route than flying south west. Thus the red-backed shrikes breeding in Portugal and northern Spain migrate to Africa by flying east across southern Europe before crossing the eastern Mediterranean and then fly south to southern and central Africa. Not all European migrants adapt such a quirky route. Other species, such as blackcaps, whitethroats *Sylvia communis*, and willow warblers *Phylloscopus trochilus*, have split their migration so that the eastern population migrates to Africa by flying around the eastern Mediterranean and the western population flies around the western Mediterranean.

One interpretation of the odd migration routes around the south-east Mediterranean is that that many species may adopt routes, not because they are currently the most sensible ones but because their distant ancestors did (Lövei 1989), for example, as a result of the location of ice age refuges. Other species, that adopt what seem to be more sensible routes by crossing at either end of the Mediterranean depending on their breeding location must have either evolved new more-direct routes or been split into a number of refuges in the last ice age.

In some cases migration systems have been suppressed. The remaining 30 néné *Branta sandvicensis* on Hawaii showed a regular pattern of migration. Like other geese, the young follow the adults for some time and the migration is presumably culturally determined. After the introduction of 1000 captive-bred birds with no knowledge of migration, the migration stopped.

There have even been a number of attempts to create new migration routes. One project is creating a new migration route for the lesser white-fronted goose *Anser erythropus* in Sweden by raising birds under barnacle geese foster parents so these migrate to the Netherlands rather than south-east Europe where they tend to be shot (Essen 1991). Peter Scott once had plans to initiate a red-breasted goose migration within Britain but this failed as the barnacle geese foster parents did not follow the anticipated route. There are currently plans in North America to train Canada geese *Branta canadensis* to adopt new migration routes by imprinting birds onto an ultralight plane and then flying from Ontario to the proposed wintering grounds in Virginia. The object is to see if this technique could be used to reintroduce rare birds, such as whooping cranes *Grus americanus,* and encourage them to migrate to the wintering areas they used to inhabit.

9.6 Partial migrants

Partial migration is the phenomenon that within a population some individuals migrate whilst others do not (Lundberg 1988; Swingland 1983). Kaitala *et al.* (1993) point out that there are three possible control mechanisms to explain partial migration. These are that it is genetically determined (Berthold 1984, 1988, 1993; Berthold *et al.* 1986, 1990, 1992; Berthold and Terrill 1991), that it is dependent upon conditions such as age or sex (Adriaensen and Dhondt 1990; Swingland 1983) or that it is an evolutionarily stable strategy (Kaitala *et al.* 1993; Lundberg 1987). However, these approaches may not be alternatives but simply different levels of explanation. In the model in Chapter 11 a genetically determined evolutionarily stable strategy is derived and it would be easy to extend this to determine the evolutionarily stable strategy for different ages and sexes.

It has been argued that partial migration should occur as a response to variation in environmental conditions (Cohen 1967; Lundberg 1987), but the model of Kaitala *et al.* (1993) shows that environmental variability itself cannot be responsible for the evolution and maintenance of partial migration and that density-dependent overwintering survival is more likely to be important.

Partial migration could be either facultative as in Swedish blue tits *Parus caeruleus* (Smith and Nilsson 1987) or obligate as in the blackcaps in southern France (Berthold 1988) or a mixture of the two. Berthold showed that when blackcaps were taken from populations that showed partial migration some individuals showed migratory restlessness while others did not. This contrasted with fully migratory populations in which all showed restlessness and sedentary populations in which none did.

Berthold *et al.* (1990) bred together those individuals which showed migratory restlessness, 83% of whose offspring showed migratory behaviour, and bred together those individuals which did not, 48% of whose offspring showed migratory behaviour. This selection experiment was repeated for a second generation and yielded 92% and 30% of migrants respectively. After a further two generations of selection on the non-migrant line the proportion of migrants had dropped to 10%. Thus a few generations of strong selection can turn a population of partial migrants into one of either migrants or non-migrants. The estimated value of the heritability was 0.6. Similar selection experiments with robins *Erithacus rubecula* and blackbirds *Turdus merula* show that their migratory behaviour also has a strong genetic component (Bierbach 1983; Schwabl 1983) and analysis of free-living stonechats *Saxicola torquata* and song sparrows *Melospiza melodia* also suggest that partial migration is genetically determined (Berthold 1984; Dhondt 1983). Berthold (1993) suggests that for bird migration all the evidence so far supports the idea of partial migration being genetically determined.

In populations of partial migrants the fitness of the two strategies is expected to be equal at equilibrium (Gauthreaux 1982; Lack 1954). About 20% of Aldabran giant tortoises *Geochelone gigantea* migrate from inland areas to the coast during the rainy season. Some individuals migrate regularly while others never migrate. At the start of the rains the principle inland food source 'tortoise turf' is dry and of little nutritional value. Tortoises moving to the coastal strip can take advantage of the perennial grass *Sporobolus virginicus*. Swingland and Lessells (1979) show that migrating individuals have a higher reproductive success than non-migrators but suffer a higher mortality, and they suggest the two strategies have the same fitness

The evidence for birds however suggests that the lifetime reproductive success is higher for residents than for migrants which contradicts the expectation of the two strategies having the same fitness. Resident robins in Belgium had higher survival and, as they could settle into breeding territories earlier, had a higher breeding success (Adriaensen and Dhondt 1990). Studies of other bird species have found the same result (Berthold 1984; Schwabl 1983). One explanation for why a proportion of the population persists in migrating is that the returns for residents vary greatly between years according to the severity of the winter. Thus in most years it is better to be a resident but in occasional severe winters it is considerably better to be a migrant. Mean fitness is an inappropriate measure as it underestimates the consequences of the occassional catastrophic year. For example a population experiencing an annual mortality of 12.5% in four out of five years and 100% (i.e. total eradication) in the fifth has the same mean mortality value as a population with a constant mortality of 30%. This can be overcome for species with an annual life cycle by using geometric fitness (Caswell 1989) or for those with longer lifespans by using the approach of Tuljapurkar (1982).

The other possibility is that the strategies are related to differences in competitive ability with the poorer competitors migrating (Gauthreaux 1978; Lundberg 1988). But as stated above, it is likely, at least in birds, that much of the partial migration is genetically determined.

9.7 Age and sex differences

In many species of birds males and females migrate different distances (Ketterson and Nolan 1983). Gauthreaux (1978, 1982) suggests that poorer competitors will leave the breeding areas to avoid competition. In most species of birds adults are dominant over juveniles and males are dominant over females. In accordance with Gauthreaux's model, juveniles usually winter further south than adults and males usually move less far than females. However the reviews of Myers (1981) and Reynolds *et al* (1986) suggest that this is a poor explanation of the distribution of migrant waders.

Two other explanations for this distribution of age and sex classes are that larger individuals might occur further north as they are better at withstanding harsh weather (Ketterson and Nolan 1976, 1983), but comparative studies of wading birds show that there are many exceptions—in many species, such as, least sandpiper *Calidris minutilla*, western sandpiper *Calidris mauri*, dunlin *Calidris alpina*, semipalmated sandpiper *Calidris pusilla*, and stilt sandpiper *Micropalama himantapus*, females are larger but do not winter further north or even winter further south than males (Myers 1981).

The final theory, which seems the most credible (Gauthreaux 1978, 1982; Myers 1981; Reynolds *et al.* 1986), is that the sex that competes most vigorously for mates arrives first on the breeding site and thus spends the winter closer to the breeding ground (Parker and Courtney 1983). In Red-necked phalaropes *Phalaropus lobatus* and Wilson's phalarope *Phalaropus tricolor* the males are solely responsible for incubation and brood-rearing and the females vigorously compete for access to males. In both species the females arrive on the breeding grounds before the males (Reynolds *et al.* 1986).

Greenberg (1980) and Ketterson and Nolan (1983) suggest competition should result in equal survivorship along a latitudinal gradient. Ketterson and Nolan (1983) showed that southern wintering dark-eyed juncos have higher survival than those wintering further north and they suggested this balanced the migration costs. However, they suggest that no single factor accounted for the differential winter distribution of dark-eyed juncos.

It seems likely that, as with partial migration, the phenomena of differential migration distances between different individuals will be largely genetically determined. The migratory behaviour in captive male and female dark-eyed juncos was compared when they were given the same photoperiod and free access to food (Holberton 1993). When they were migrating the two showed equal intensity of migratory behaviour but on average females showed migratory behaviour for 21 more nights.

9.8 Summary

The approach used for studying the use of patches within sites can be used for determining the choice of sites. It is then possible to determine evolutionarily stable migration routes which incorporate the competition caused by others of the same or different breeding populations.

Some species have changed their migration routes over recent decades. Many German blackcaps have shifted from migrating to the south-west Mediterranean to over wintering in Britain, red-breasted geese have shifted migration routes in response to habitat change, and introduced canada geese have developed migration routes in Britain. This suggests that migratory behaviour can respond rapidly to changing circumstances.

There is however other evidence that some species have been slow in changing routes. Many of these adopt apparently unfavourable routes and one reasonable interpretation is that these routes are a consequence of their history, for example as originally adopted in relation to ice age refugia.

Males and females and adults and juveniles often have different migration routes. Although it is possible that this is related to differences in competitive ability, the evidence suggests that other factors, such as the need for the sex that competes for mates to return early to the breeding ground may be more important.

10

Applied problems

10.1 Introduction

Previous chapters have explained in general terms how the distribution of consumers in relation to the food population can be understood. Although this approach can be used for examining the general principles applying to any species, the details of the natural history will influence the outcome. Of particular importance is unravelling the factors affecting habitat preferences. The two aims of this chapter are firstly to describe how to develop models of a given species in a given area and secondly to show how these models can be used to tackle applied problems. The actual problems involve the two following predictions: the consequences of increases in the number of competitors and changes in habitat management on the species of conservation interest; and the extent of future agricultural loss resulting from species whose populations are increasing.

10.2 Models of specific systems

I will illustrate how models of specific systems can be created by giving three examples, all of which are wintering populations of migratory species of geese. These models relate to a population of bean geese *Anser fabalis* in the Yare Valley, eastern England, which are of national conservation interest; pale-bellied brent geese *Branta bernicla hrota* at Lindisfarne, north-east England, which comprise the majority of the Spitzbergen wintering population; and dark-bellied brent geese *Branta bernicla bernicla* on the north Norfolk coast in eastern England whose population is increasing and coming into conflict with farmers. The models described here are all based on the depletion models described in Chapter 2.

The essential components for creating any such model are to divide the study area into patches and then determine:

(1) the resource density in each of the patches;

(2) the factors affecting patch choice by the consumers;

(3) whether the resource increases, either due to growth or reproduction, and if so at what rate;

(4) the rate at which each consumer depletes the resource;

(5) the level of resources necessary to sustain the consumers and below which they will starve or emigrate;

(6) the seasonal changes in the number of consumers.

To apply the general models to specific situations, details of the species' ecology need to be considered. In the three examples considered here, it was necessary to incorporate factors such as the role of competing species and the movements of the tide. For other studies there would undoubtedly be a requirement to incorporate different details. I will describe all three models first and then describe the results.

10.2.1 *Bean geese*

The Yare Valley contains the only regular wintering flock of bean geese in England. Part of this valley, known as Buckenham Marsh, used to be the major site. The model was created to examine why the population has moved elsewhere in the Yare Valley, the consequences of habitat management, and the nature of competition with wigeon *Anas penelope*. The ecology of both the wigeon and bean geese has been studied by Allport (1991) and I used this information to create the model described here (Sutherland and Allport 1994). The approach illustrated in Fig. 10.1 was used to determine the number of bean geese that can be sustained in the fields at the Buckenham study site.

The grass productivity was determined by measuring marked tillers in swards of different heights, placing an exclosure around them and remeasuring them after two weeks (Peacock 1975). This quantified the well-known phenomenon that plants grow faster both at higher temperatures and when the sward is short:

$$\frac{\text{productivity}}{\text{(in gm}^{-2})} = \left\{ 0.029 \times \frac{\text{temperature}}{\text{(in °C)}} \right\} - \left\{ 0.0008 \times \frac{\text{biomass}}{\text{(in gm}^{-2})} \right\} + 0.038 \qquad (10.1)$$

A standard way of assessing wildfowl numbers is to count the droppings as these are produced at a reasonably constant rate (Owen 1971). The biomass preferences of the wigeon and bean geese were determined by relating the numbers of birds (by counting their droppings) to sward height. This was carried out within single fields; as the model describes the number of birds on each of the 40 fields the model is not simply redescribing the data.

For the simulations the winter was divided into 26 weeks and for each week the number of wigeon, number of bean geese, and temperature were added.

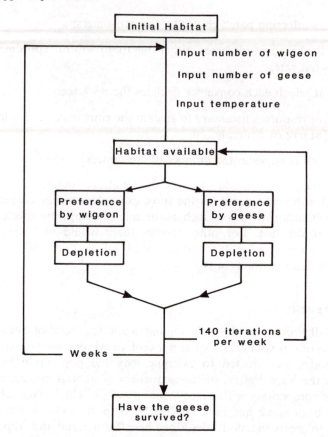

Fig. 10.1 The framework for modelling the grazing caused by bean geese and wigeon.

Each week was divided into 140 time periods and for each time period, the field each species most prefers and the resulting depletion was determined. For both wigeon and bean geese, daily depletion rates and seasonal changes in population size were incorporated. Thus as time elapses, the distribution of both species changes as depletion and productivity alter the biomass in each field. The threshold grass biomass that bean geese require to obtain sufficient intake was determined. It was then possible to determine the conditions under which the geese could survive through the winter.

Figure 10.2 shows the relationships between the number of droppings and grass biomass for both wigeon and bean geese. Bean geese selected the taller swards and this depletion is straightforward to model. Wigeon selected intermediate length swards; they clearly avoided short swards, presumably because they obtain a lower intake, and also avoided long swards, perhaps because it is harder to see predators when in long grass or because longer swards are less palatable. With a preference for intermediate swards depletion

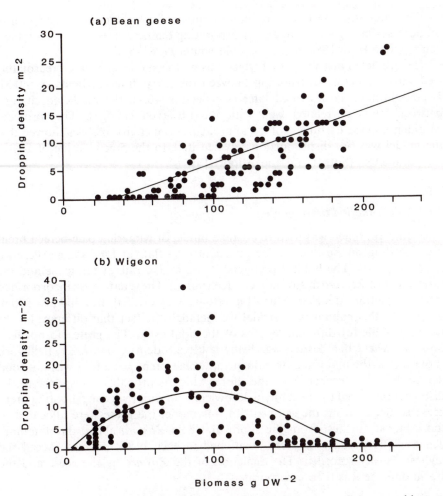

Fig. 10.2 The relationship between the number of droppings and the grass biomass in grams dry weight per m² for (a) bean geese and (b) wigeon. (After Sutherland and Allport 1994.)

becomes more complex. From a wigeon's perspective, a field with a high biomass will become more attractive as it is depleted to nearer the preferred height. Once the sward is at the preferred height any further depletion will then make it less attractive.

The aim of the simulation was to predict the number of bean geese that could persist under certain conditions. This number was expressed as a multiple of the current population. The technique used was to determine the increase in bean geese that could occur before the grass is depleted below the threshold necessary for survival. As the bean geese at Buckenham are restricted by disturbance to a set of five fields, they were similarly restricted in the model.

The simulations were carried out for a range of initial biomasses and a range of increases in wigeon numbers, incorporating temperatures for both a typical winter (data from 1987–88) and a cold winter (1981–82).

The predicted distribution of wigeon using this model correlated reasonably well with the actual distribution between the forty different fields ($r_s = 0.31$, $P < 0.001$) and, as described later (see Section 10.4), the predicted changes between years correlated well with the observed changes. This provides confidence in the usefulness of the approach. A sensitivity analysis showed that the model was not disproportionately sensitive to the precise value of any of the parameters (Sutherland and Allport 1994).

10.2.2 *Pale-bellied brent geese*

The same approach was used to create a model of wintering pale-bellied brent geese feeding on *Zostera* on the intertidal mud-flats at Lindisfarne (Percival *et al.*, in press). The habitat preferences and intake rate of the geese and the rate of loss of *Zostera* in storms were determined. These data were incorporated into a depletion model in which the estuary was divided into 500 m × 500 m blocks. As this habitat is intertidal we included the fact that different blocks were available for different lengths of the tidal cycle. The geese did not feed on areas where the *Zostera* was below a certain density. However, fieldwork showed that the major factor influencing where brent geese feed was not food density but the distance from the shore, which is probably a response to the disturbance caused by the many wildfowlers who shoot wigeon (shooting brent geese is illegal) from the sea wall. In this model the geese were allocated to those blocks furthest from the shore that held *Zostera* above the minimum density. The number of geese positioned in each block did not exceed the highest density recorded. The depletion of the *Zostera* by the geese and the storm damage was then incorporated.

10.2.3 *Dark-bellied brent geese*

One of the main objectives of studying brent geese in north Norfolk, England, mentioned in Chapter 3 was to consider the relationship between the total number of geese in the site and the extent of feeding on agricultural land. During a season, the geese feed on intertidal algae, then salt-marsh plants, and finally on agricultural crops inland. The modelling was approached in two ways. One was to produce detailed models of intertidal algae and salt-marsh, as in the previous two examples, and combine them to predict the point at which birds feed inland (Rowcliffe 1994). The second approach was less sophisticated but gave similar results. This was to calculate the number of bird-days spent on the intertidal habitats and assume this was independent of the total population. Thus, if the population doubles, each goose can only spend half as many days feeding on the intertidal habitats. As the population increases they deplete these intertidal resources more rapidly and move inland earlier.

These models of bean geese, pale-bellied brent geese, and dark-bellied brent geese were used to explore three different conservation issues. They were also used to consider the consequences of habitat loss as described in the next chapter.

10.3 Predicting the consequences of changes in the number of competitors

For both the bean geese at Buckenham and the pale-bellied brent geese at Lindisfarne it has been suggested that numbers have declined due to increases in the numbers of wigeon. Whether this is likely can be explored using these models. The numbers of wigeon have been steadily increasing in the Yare Valley (Fig. 10.3) and the model shows that this has marked consequences for the predicted number of bean geese that can be sustained (Fig. 10.4). Increases in the numbers of wigeon lead to considerable decreases in the numbers of bean

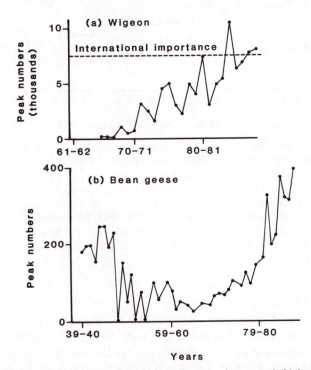

Fig. 10.3 The annual number of overwintering (a) wigeon and (b) bean geese in the Yare Valley each winter. The internationally important population size (1% of the north-west European population) of wigeon is shown. (After Sutherland and Allport 1994.)

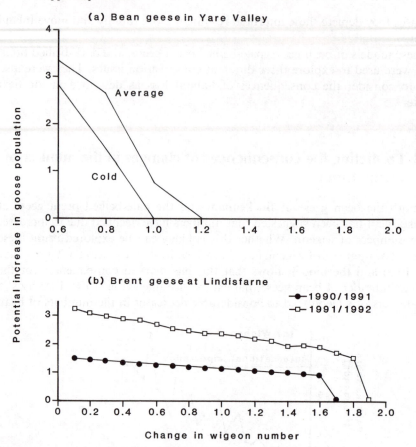

Fig. 10.4 The predicted relationship between the number of wigeon and the size of goose population that can be sustained. Thus a value of 1 means that the entire current English population can be sustained at the site, 0.5 means half the population could be sustained, and 2 means that twice the current population could be sustained. (a) the proportion of the English bean goose population that could be sustained at Buckenham for cold and average winters (Sutherland and Allport 1994). (b) The proportion of the English pale-bellied brent geese that can be sustained at Lindisfarne in relation to the numbers of wigeon. (From Percival *et al.*, in press.)

geese that can be sustained as a result of the direct competition between the two species. Indeed, the model shows that with the recent management, and the current populations of wigeon, the Yare Valley bean goose population could not be sustained at Buckenham, the traditional major wintering area. This is, in fact, what happened and in recent years the geese have spent most of the winter elsewhere in the Yare Valley.

The study shows that one way of improving Buckenham for bean geese is to reduce the numbers of wigeon present (Fig. 10.4(a)). Scaring wigeon was attempted at one time but it failed as they simply flew to adjacent fields. Furthermore, although in Britain wigeon are common and bean geese are scarce, from a European perspective the Yare Valley holds internationally important numbers of wigeon (i.e. over 1% of the north-west European population) but internationally trivial numbers of bean geese!

At Lindisfarne large numbers of wigeon are also present in the autumn and there is a popular belief that this is detrimental to the pale-bellied brent geese. However, in this case, the model shows that any increase in wigeon numbers would have a negligible effect on the numbers of brent geese that could be sustained (see Fig. 10.7). The two years studied differed in the *Zostera* density and the number of wigeon present. This result contrasts with the results of the bean goose model. This is because the peak numbers of wigeon are present before most of the natural loss of the *Zostera* in the autumn storms. Thus, much of the biomass removed by wigeon would be lost anyway.

10.4 Predicting the consequences of habitat change

The habitat available to a population may change as a consequence of changes in management or other external forces such as global warming. The models described here can be used to explore such changes. The model of bean geese was run with a number of different grazing regimes. The sward is usually higher after cattle grazing than sheep grazing. The pattern and intensity of stock grazing has varied between years. At Buckenham there is a clean grazing system which means that an area grazed by cattle one year was grazed by sheep the next (Fig. 10.5). The actual populations of bean geese varied in the manner expected from the model output (Fig. 10.6). In 1987–88 the site was managed in a manner sympathetic to the geese by reducing the grazing in a number of fields. The model predicted that this should be much more suitable for the geese and this was the case (Fig. 10.6).

Over the past decade, the increase in both sheep grazing and the wigeon population at Buckenham has been matched by a series of unusually warm winters. The model suggests that the wigeon population would encounter problems if faced with a cold winter. Thus, in the cold winter of 1990–91 the wigeon depleted both the study site of Buckenham and the adjacent site of Cantley and moved to new sites within the Yare Valley. Such alternative sites are likely to be more prone to disturbance and shooting.

This model was used to explore alternative grazing options that showed the optimal management condition for the bean geese. Both the model output and the results of change in management in 1987–88 (Fig. 10.6) show that the management of this site can be improved to the benefit of the geese. The straightforward solution is to increase the area under cattle grazing, or to ensure that, prior to the autumn arrival of the geese, the sheep are removed from the fields preferred by the geese in time for the sward to recover to a sufficient height to sustain the population of bean geese throughout the winter. After

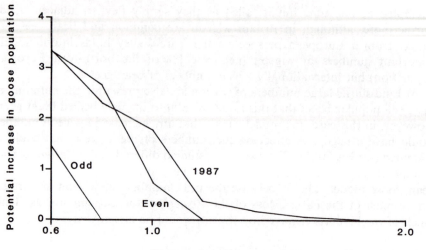

Fig. 10.5 The fraction of the current English bean goose population that could be sustained at Buckenham under three different grazing regimes. The management carried out on odd years (largely sheep grazing), the management carried out on even years (largely cattle grazing), and the management carried out in 1987 (stock removed in the autumn from the five main fields used by the bean geese so allowing the sward to recover).

this model was created, the Royal Society for the Protection of Birds bought Buckenham in 1992 and managed it in the manner that the model predicted would be optimal. This was a success, with the highest number of geese for a decade.

The field work at Lindisfarne showed that there was a considerable loss of *Zostera* each autumn due to storm damage. The severity of the loss varies between years. The model shows (Fig. 10.7) that this is an important factor and that an increase in storm loss (as some suggest may result from global warming) would affect the geese.

10.5 Predicting the consequences of increases in population

As a result of protection from shooting within its winter range, the world population of dark-bellied brent geese has increased (Ebbinge 1991). The consequences of an increase in the numbers of brent geese wintering on the north Norfolk coast has been explored by Vickery *et al.* (in press). On this site brent geese numbers have increased from 245 in 1955 to a population of about 4500 in the early 1990s.

The geese currently use a range of habitats. They feed on algae early in the winter, then shift to feed on the salt-marsh, and then move inland to feed on crops later in the winter (see Chapter 2). With an almost 20-fold increase in

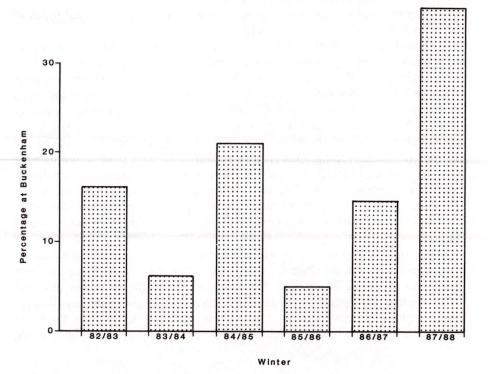

Fig. 10.6 The percentage of the Yare Valley bean goose flock using Buckenham each winter. (After Sutherland and Allport 1994.)

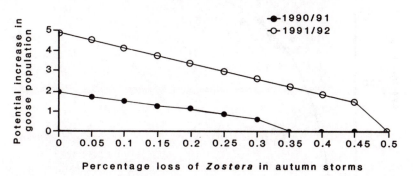

Fig. 10.7 The numbers of pale-bellied brent geese that can be sustained at Lindisfarne in relation to the percentage storm loss. (From Percival *et al.*, in press.)

the goose population, and the resulting enhanced depletion, the geese can be expected to switch habitats earlier. This is what has actually happened—the date of the switch from intertidal mud to salt-marsh has changed from late

February to late October, and pasture and cereals are now used for much of the winter where previously they were not used at all.

The brent goose population has increased at about 7.4% per annum and all the evidence points to the brent goose population increasing further (Summers and Underhill 1991). It seems inevitable that the rate of depletion on both algae and salt-marsh will increase and the birds will thus switch inland even earlier. Figure 10.8 shows the theoretical relationship between the number of birds in North Norfolk and the duration of inland feeding. An increase in the goose population will therefore result in a disproportionate increase in the amount of inland feeding. This graph also shows the data points for a number of years between 1955–58 and 1983–92 which illustrate that once the population exceeded a certain threshold the extent of inland feeding increased markedly. This was attributed by some to cultural learning but it is clear from this graph that cultural learning is not necessary to produce a sharp increase in inland feeding.

Although brent geese are considered by some to be a pest species such an assessment very much depends upon the perspective. The financial consequences of inland feeding can be calculated. From the reduction in cereal yield, the loss to a farmer is approximately 3.9 pence per goose per day (Vickery *et al.* 1994). Not surprisingly, farmers are reluctant to tolerate flocks of thousands of geese on their land. The twist to the calculations is that cereal

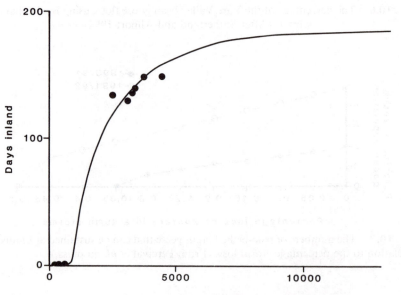

Fig. 10.8 The theoretical (line) and actual (points) relationship between brent goose numbers and the population feeding inland. The points show data from different winters. (After Vickery *et al.*, in press.)

production is heavily subsidized by the taxpayer within the European Union and from the perspective of the taxpayer, each goose saves 1.9 pence per day in reducing agricultural surpluses! One sensible solution is to provide compensation from the same funding that pays for set-aside. We can predict that the amount of goose grazing inland will increase both as the population increases, at a probable average rate of 7.4% per annum, and as they consequently graze inland for longer.

10.6 Summary

Models can be used to describe given species in specific areas by incorporating field data. This approach is used to consider the consequences of habitat management, the changes in numbers of competitors, and the consequences of an increasing population that causes agricultural damage. The three following examples are given: bean geese in the Yare Valley, which are of conservation interest and models have shown how they stopped using their main wintering area as a consequence of increased competition with wigeon and inappropriate stock management; pale-bellied brent geese, which also feed with wigeon but in this case the modelling showed that the wigeon are not a real problem; and dark-bellied brent geese, which are causing conflict with farmers through inland feeding, of which the model shows the likely continued increase. The financial consequences of inland feeding by dark-bellied brent geese is also estimated.

11

Habitat loss

11.1 Introduction

There is widespread concern over the damage that humans are inflicting upon the environment. One of the consequences of this is a reduction in area of various habitats. Further habitat loss may be expected due to processes such as industrial development, agricultural intensification, deforestation, and sea level rise.

There is a need to be able to predict the consequences of such habitat loss for species of conservation concern. For example, in public enquiries into development proposals, it is common for conservationists to argue that a loss of habitat will result in a reduction in the total population of associated species of conservation concern while developers often argue that the displaced animals will survive by simply moving elsewhere.

The objective of this chapter is to consider the impact of habitat loss upon populations. This will be examined firstly by considering the consequences for the population within a single site, then the consequences for migratory populations, then finally incorporating the genetics of migration to consider how rapidly species may respond to habitat loss.

11.2 Consequences within single sites

I described in Chapter 3 how models incorporating depletion can be used to predict the population that can be sustained by a site. We can then use this model to show the consequences of habitat loss within a site for the population that can be sustained by that site. Figure 3.2 shows how the sustainable population can be determined for ten patches varying in resource density. Figure 11.1 shows the consequences of removing patches. Two extreme cases are considered: either the best patches or the worst patches are removed first. This approach is suitable for examining the consequences of habitat loss for the maximum population that can be sustained.

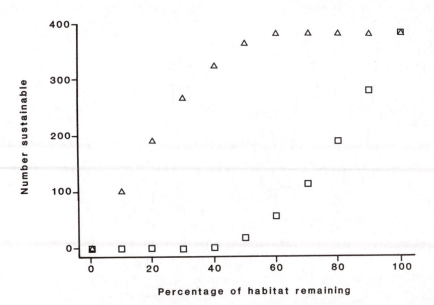

Fig. 11.1 The consequences of habitat loss on the population that a site can sustain assuming that depletion is the sole process. The parameters are the same as for Fig. 3.2. □ = poorest patches removed first, △ = best patches removed first. (From Sutherland and Anderson 1993.)

The number of wading birds in different patches has been shown to depend upon both the patch size and resource density (Bryant 1979; Goss-Custard *et al.* 1977). If part of a patch is removed then the species will be expected to redistribute between the remaining patches in the site. Thus the population within the damaged patch may decline (as they have moved elsewhere) but this is not evidence that the population as a whole will decline.

11.3 Migratory populations

It is not straightforward to consider the consequences of habitat loss for migratory species. The loss of one wintering or breeding site may result in increased numbers using the remaining sites, which may in turn lead to a lower equilibrium population size due to increased depletion and interference. The consequences of habitat loss may be explored by extending the migration model described in Chapter 8. In this model there is a range of wintering and breeding sites for which the resulting evolutionarily stable migration routes can be determined (Fig. 11.2). It is then possible to explore the consequences of habitat loss in this system by removing all or much of one of the sites and recalculating the evolutionarily stable migration routes and the population size once one of the sites has been damaged or destroyed (Sutherland and Dolman 1994). The

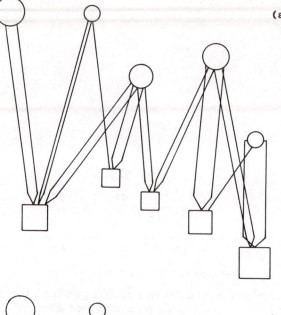

(a)

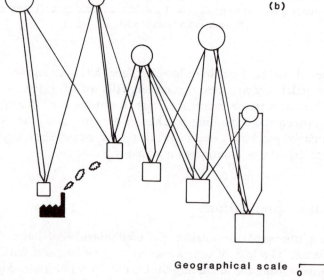

(b)

Geographical scale ⊢————⊣
0   500km

Post breeding numbers using ⊢—⊣
route scaled by arrow width 0 2000

Fig. 11.2 Consequence on the remaining migration system of removing part of a site. Each site consists of five equal sized patches differing in prey density. (a) is the same as Fig. 9.1. (b) used the same parameters but with 75% of all patches in one wintering site removed. (From Sutherland and Dolman 1994.)

problem of whether the population will shift routes is discussed in Section 11.4. Obviously many or all of the individuals no longer migrate to the damaged or destroyed site once the new migration system has been established but migrate elsewhere. A consequence of this is an increase in density in the remaining sites resulting in an increase in density-dependent starvation such that the total population declines. In the case illustrated here the decline in the total population would be 21%.

The consequences of habitat loss may also be subtle and widespread. Figure 11.2. shows how the damage to one site may affect the wintering populations, breeding populations, and migratory routes over a wide area. Adjacent breeding populations may be affected even though no individuals ever used the now destroyed wintering site. Individuals that would have used the now destroyed site instead winter on other sites and the increased levels of interference and depletion may result in higher starvation rates. It is even possible that these adjacent breeding populations may shift their migration route to avoid this increased competition so that further populations are affected. Thus, at least in theory, destroying wintering sites may affect populations that had no link with the site that was lost.

This model describes cases in which habitat loss takes place suddenly. It is probably much more common for parts of sites to be gradually lost. The same approach of considering the new evolutionarily stable migration route and population size can be used for considering the consequences for the total population of piecemeal habitat loss. Figure 11.3 shows the consequences of

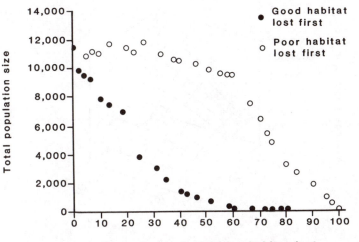

Fig. 11.3 The relationship between piecemeal habitat loss and the equilibrium population size. The parameters are the same as for Fig. 11.2 but with single patches removed from within wintering sites. In one simulation the patches holding the most consumers were removed first, while in the other the patches holding the fewest consumers were removed first. (From Sutherland and Dolman 1994.)

removing a succession of single patches in the wintering areas. The simulation was carried out to show the extreme scenarios of either the patches holding the fewest individuals being removed first or patches holding the most being removed first.

Exactly the same analysis could be carried out to show the consequences of the loss of breeding sites for the population size. The general case is shown in Fig. 11.4. The impact of losing a given area of wintering habitat depends upon the relative extent of the two habitats. If populations show a time-lag in responding to the range of sites remaining then the short-term decline may be greater than predicted here (see Chapter 9 and Section 11.4 below).

11.4 The evolutionary response to habitat loss

The approach adopted so far has been to consider the new evolutionarily stable migration routes and equilibrium population size that would occur after some habitat had been lost. In reality there is likely to be a time-lag before new migration routes are established and in this section I will consider the consequences of a delay in response on the population size.

The consequences of habitat loss on migration routes and population size were explored using a simple genetic model (Dolman and Sutherland 1995). Migration routes may be determined culturally, as in geese and swans with extended parental care, or genetically, as is probably the case for most

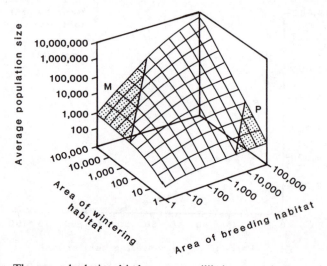

Fig. 11.4 The general relationship between equilibrium population size and the extent of wintering and breeding habitat. In shaded area M a loss of wintering area has a negligible effect on population size while a loss in breeding area has a considerable effect, while in area P a loss of breeding habitat will have a negligible effect although a loss of wintering area has a considerable effect. Elsewhere, a loss of either wintering or breeding habitat will result in a population decline. (From Dolman and Sutherland 1995.)

passerines. The only species in which the factors determining migration route have so far been studied in detail is the blackcap and in this case the genetics influencing the direction and duration of migration seem relatively simple (Berthold 1988; Berthold *et al.* 1992). There is also evidence from studies of fish of a genetic component to migratory behaviour (Hansen and Jonsson 1991; Jonsson 1982; Smith 1985).

Figure 11.5 shows a simplified version of the previous model with just two breeding areas A and B and two wintering areas 1 and 2. All individuals from B winter in area 2. Most individuals from A winter in area 1 but a proportion winter in area 2. In the various simulations the proportion of individuals breeding in A that winter in area 2 is varied by adjusting the relative sizes of the wintering and breeding areas. We assume that the migration route taken by individuals from A is determined by a single locus and that moving to area 1 is dominant. Thus both homozygotes MM and heterogygotes Mm winter in area 1 while homozygotes mm winter in area 2. This model is, of course simplistic, and it is likely that migratory behaviour will involve many loci and alleles. However, it should illustrate the general processes.

This model can then be used to explore the consequences of losing area 1 for the population breeding in A. With the values of parameters used in this simulation the size of the population breeding in area A were 480 000 before the loss of area 1 and 200 000 after that area was lost and a new migration route was established. However, the population may initially drop even lower

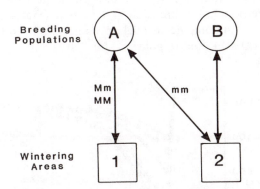

Fig. 11.5 The simple genetic model. All individuals from B migrate to 2. From A, those individuals of genotype MM and Mm migrate to 1 while those of mm migrate to 2. The consequences of losing 1 are then explored. The model is the same structure as Fig. 9.1. The wintering areas are initially equal in size, both 1000 km². Density-dependent breeding parameters in both sites are $a = 0.1$, $b = 0.3$. Annual per capita fecundity in the absence of density dependence is 2. Density independent mortality of 10% per annum acts on the post-breeding population. The winter areas consist of 10 patches whose initial prey density ranges from 10 to 50 items m⁻². Handling time $= 5$ s. Quest constant $= 0.001$ m² s⁻¹. Threshold intake for survival $= 0.04$ items s⁻¹. Coefficient of interference, $m = 0.2$. There are 100 competitive phenotypes, normally distributed with a mean of 10 and a variance of 4.

as shown in Fig. 11.6. The delay in reaching the new equilibrium depends upon the initial frequency of the m allele. If m is initially frequent, say 20% of the total, then the population initially drops to about 10 000 and then takes about 50 generations to increase to the new equilibrium. However, if the allele is initially scarce, then the population may plummet and take even longer to recover. In the case of the allele initially occurring in only 0.5% of the population then the population drops to about 10 individuals for many generations and thus in reality would probably go extinct due to stochastic demographic events. This shows that any recovery response to environmental change may have a long time-lag and suggests that in some cases populations could even fail to respond to dramatic changes although the selection pressure may be very large.

Instead of considering area 1 as disappearing rapidly, as might happen if the entire area was developed, we can examine the consequences of gradual loss, for example from sea level rise or piecemeal developments (Fig. 11.7). As the habitat is lost, there is, of course, strong selection for the m allele. If the allele for migration to area 2 is moderately abundant (18.5% in this example), then, under the strong selection resulting from the habitat loss, the m allele spreads to fixation as the habitat is lost. As a consequence of the habitat loss the population declines, but it more or less tracks the extent of habitat loss.

If the m allele for migration to site 2 is initially rarer (say 1%) then the consequences of piecemeal loss are very different (Fig. 11.8). Although again there is strong selection for wintering in site 2, homozygote recessives are rare in each generation so that selection will have little effect on the abundance of the m allele. The population breeding in area A reaches such low levels it is likely to go extinct before the m allele becomes sufficiently frequent to be regularly expressed as a homozygote.

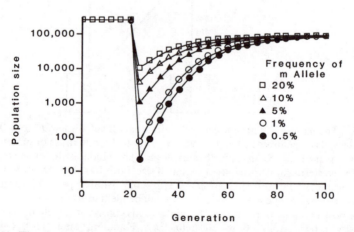

Fig. 11.6 The change in population size of A over time after the sudden loss of winter site 1. A range of initial frequencies of the m allele are shown. Parameters as in Fig. 11.5. (From P.M. Dolman and W.J. Sutherland, unpublished.)

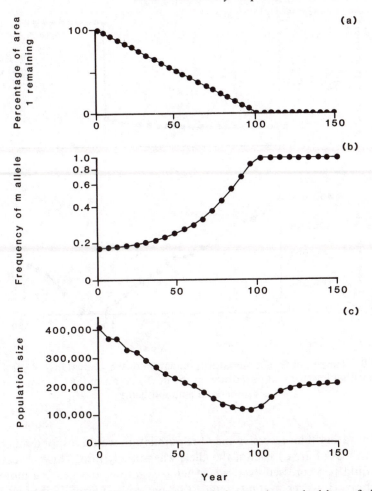

Fig. 11.7 The response of a migrating population to the gradual loss of site 1. (a) shows the amount of site 1 remaining; (b) shows the frequency of the m allele in the population; and (c) shows the size of the population. Parameters as in Fig. 11.5. (From P.M. Dolman and W.J. Sutherland, unpublished.)

The extent to which populations can track the pattern of habitat loss will, of course, also depend upon the rate at which the loss occurs. Figure 11.9 shows four different rates at which the area is lost over periods varying between 5 and 100 generations. With a slow loss of habitat, the population can evolve and change migration routes so that the population tracks the habitat loss. However with more-rapid habitat loss, the population is unable to evolve at the necessary rate and, as a consequence, the population declines before reaching the new equilibrium.

This model is of course simplistic and the outcome would vary with the precise details of the genetics. If the allele to migrate from A to area 2 was

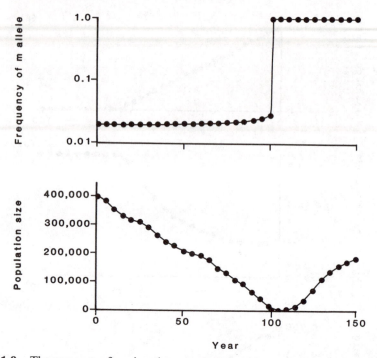

Fig. 11.8 The response of a migrating population to the gradual loss of site 1 in Fig. 11.7 but with a lower initial proportion of the m allele. (From P.M. Dolman and W.J. Sutherland, unpublished.)

dominant rather than recessive then the population would respond much more rapidly to loss of area 1 even if the allele was initially rare. This is because the allele would be expressed even when heterozygotous and thus be more open to selection. However mutants tend to be recessive (Fisher 1958) and it may thus be that mutants for novel migration routes also tend to be recessive.

Some blackcaps breeding in Germany have altered their migration route in the last few decades. They all used to fly south-west to the Mediterranean but many now fly north-west to Britain (see Chapter 8). The analysis here suggests that such a rapid increase of rare alleles is impossible! One possible explanation is that the allele for migrating north-west is dominant—but the evidence contradicts this (Helbig *et al.* 1994).

A better explanation for this rapid change is that assortative mating is important. Fig. 11.10 shows the consequences of incorporating assortative mating into the model by similar phenotypes pairing with each other. Whether individuals of the same phenotype were homozygous or heterozygous had no effect. The greater the assortative mating, the more rapidly the population responds. A result of such assortative mating is that recessive homozygotes will be more frequent than expected from the Hardy–Weinberg equilibrium and thus gene frequencies will change much more rapidly as a result of selection.

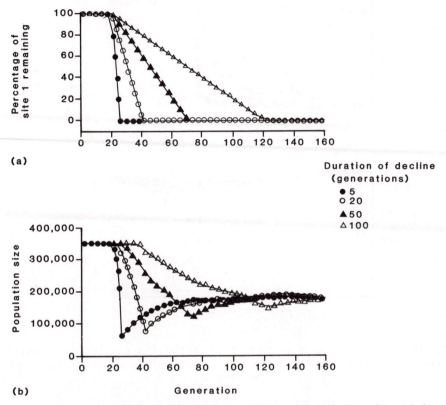

Fig. 11.9 The response of the populations to different rates of habitat loss. (a) shows the different rates of loss and (b) shows the responses of these populations. Parameters as in Fig. 11.5. (From P.M. Dolman and W.J. Sutherland, unpublished.)

Captive blackcaps given the photoperiod they would experience in Britain started their migratory activity earlier than did those that were given the photoperiod that would be experienced by those in their usual wintering site in Spain (Terrill 1990). This implies that the birds wintering in Britain may return to the breeding grounds about ten days earlier. This is likely to result in assortative mating as the earlier-arriving individuals pair with each other. There is also some field evidence that birds wintering in Britain return earlier. Berthold (1995) arranged for the first migrants returning to Germany to be captured and retained until the autumn—once tested they migrated in a direction towards Britain. Those returning later migrated towards the south-west. A further possible selective advantage for migration to Britain is that the first birds to return are more likely to obtain territories and more likely to obtain better territories.

Such assortitive mating could occur in a range of situations. Sedentary blackbirds in Germany tend to breed close to towns while migrants tend to

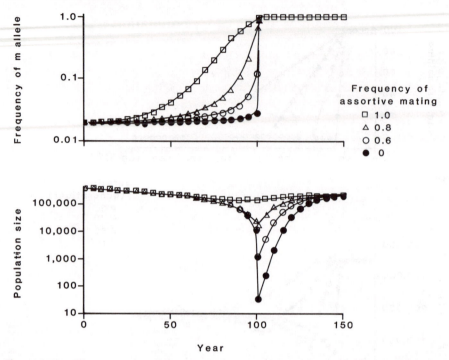

Fig. 11.10 The response of the population to habitat loss with different degrees of assortative mating. Parameters as in Fig. 11.7. (From P.M. Dolman and W.J. Sutherland, unpublished.)

breed in woods resulting in birds of similar behaviour tending to breed together (Schwabl 1983).

The ability to change migration routes is obviously different for species in which the migration pattern is culturally determined. Cultural ideas may be passed on from parents to offspring in a manner similar to genetical inheritance and this is referred to as vertical transmission. Such a process would be relatively easy to model using a similar approach to the genetic model described here. However a major complication is that ideas may also be transmitted between unrelated offspring (horizontal transmission) or between parents and unrelated offspring (oblique transmission). Genetic models are clearly inappropriate and the best approach seems to be to use epidemiological models as these readily incorporate oblique and horizontal transmission (Cavalli-Sforza and Feldman 1981). There are records of species that learn migration routes from others yet have changed routes (as described in Section 8.3) but further work is necessary before we can understand the mechanisms and constraints.

11.5 Observed changes in migratory populations

Although the extent of many habitats has declined, much of the interest in the effect of habitat loss upon populations of migratory species has focused on migratory populations of waders on estuaries. The intertidal habitat in estuaries in Britain has disappeared at the rate of 0.2–0.7% per year (Davidson *et al.* 1991) and in the United States wetlands have disappeared at 0.2–0.5% per year (Gosselink and Aumann 1980). In more than half of the estuaries in Britain there are currently proposals for developments such as marinas and barrage schemes. In some cases the justification for such destruction seems particularly slim. For example, the estuary of Cardiff Bay, Wales, seems likely to be completely destroyed by the building of a barrage largely on the grounds that the estuary was considered aesthetically displeasing with the tide coming in and out, covering and exposing the mud-flats.

There have been a number of studies examining the consequences of habitat loss. In the Tees estuary, England, 260 ha out of 400 ha were reclaimed in 1974. The wading birds declined in proportion to the area lost or at an even greater rate (Evans *et al.* 1979). Of the Danish Wadden Sea 1100 ha were reclaimed in 1980 and of the 12 wader species found there, eight declined by 85% and the most important duck species declined by 60% (Laursen *et al.* 1983). There has been a long history of habitat loss on the Forth Estuary, Scotland and when 20% of the feeding area was lost to land reclaimation, dunlin, bar-tailed godwit, oystercatcher, turnstone, and knot declined by over 20%, but redshank and wigeon numbers did not change, while curlew and shelduck increased (Bryant 1987; McClusky *et al.* 1992). Goss-Custard and Moser (1988) showed that the rate of decline of dunlin on different estuaries was proportional to the extent of encroachment of *Spartina anglica*.

The most detailed example is Meire's study of the impact of the tidal barrage constructed on the Oosterschelde estuary, the Netherlands, which resulted in a 33% decrease in intertidal area of the estuary. Prior to the completion of the barrage, Meire and Kuijken (1987) made predictions that the wader populations would decline substantially in the Oosterschelde. The number of waders wintering in the estuary decreased although the number of ducks increased. There was no evidence that the waders simply fed at higher densities in the remaining mud-flats (Schekkermann *et al.* in press). In this study there were three severe winters just prior to the barrage being completed in which many birds and benthic invertebrates died. This illustrates the problems of such studies. They are unreplicated and the changes in populations are easiest to interpret if the environment stays constant—which of course it rarely does.

Such detailed studies of specific developments are very useful but are often insufficient to answer the main question as to what are the consequences of habitat loss on the total population. Birds are known to distribute themselves between estuaries according to the estuary size and prey abundance (Goss-Custard *et al.* 1977) and it is thus only to be expected that reducing the size of an estuary will eventually lead to a reduction in the number of birds using the estuary . Whether the displaced birds settle elsewhere or whether the total population declines is still unanswered. It is, however, very interesting to document the speed and manner in which the habitat loss affects local populations.

Some declines in migrant birds are clearly due to habitat loss in the breeding grounds. The breeding populations of many wading birds have declined in Europe due to the drainage and reseeding of pastures (Batten *et al.* 1990). There may, however, be a time-lag in the response to breeding habitat loss. In Oregon, as part of a programme to plant wheatgrass *Agropyron cristatum*, an area of 16 km^2 was sprayed with the herbicide 2,4,dichlorophenoxyacetic acid and the sagebrush shrubs were broken down and removed. Sage sparrows are strongly dependent on sagebrush yet the numbers did not change even when the coverage dropped from 27% to 4% (Weins 1985). This lack of response was attributed to site tenacity as within the study period the adults persisted; Hildén (1965) documents similar cases. It seems that as with winter sites described earlier (see Section 9.5), adults tend to persist within an area while juveniles settle according to the suitability of the particular area. Thus breeding habitat deterioration may result in reduced breeding success and a lack of juvenile settlement but the breeding population may only slowly decline as the adults die and are not replaced.

As well as loss of breeding or wintering areas the prevention of migration may also be a problem. Fences built across much of Botswana to prevent the spread of foot and mouth disease from buffalo *Synceros caffer* to cattle *Bos primigenius*, have resulted in considerable population decreases in mammals unable to migrate. In one case 50 000 wildebeest *Connochaetes gnou* died. Building dams across rivers is similarly well known to reduce populations of salmon *Salmo salar* by restricting their migration.

Terborgh (1990) collated evidence showing that there have been declines in the populations of a number of migratory species in North America. After analysing the various possibilities, Terborgh concluded that the most likely explanation was deforestation in the Neotropics. Terborgh's pioneering book has encouraged a number of other studies whose results are often contradictory and it is not clear if there really have been any population declines (Böhning-Gaese *et al.* 1993; Collins and Wendt 1989; Johnston and Hagan 1992; Robbins *et al.* 1979; Sauer and Doege 1990). These discrepancies between studies may in part be a consequence of the different statistical techniques used by the different studies (Thomas and Martin, in press). The censuses of eastern temperate forests collated by Johnston and Hagan (1992) seem to provide the most convincing evidence for declines in songbirds.

If there are declines it is not clear whether these are due to habitat loss in the neotropics (Terborgh 1990), habitat fragmentation in the breeding grounds leading to increased predation along forest edges (Andrén and Angelstam 1988; Böhning-Gaese *et al.* 1993; Wilcove 1985), or increases in the parasitic brown-headed cowbirds (Robinson *et al.* 1993).

As an example of the problems of interpretation, the cerulean warbler *Dendroica cerulea* has declined by 3.4% per year between 1966 and 1987 (Robbins *et al.* 1992). It has both specific breeding requirements (mature floodplain forest with tall trees) and wintering requirements (primary humid evergreen forest along a narrow elevational zone at the base of the Andes). This wintering habitat is among the most intensively logged and cultivated regions of the Neotropics. But the breeding habitat is also being lost and it is difficult to partition out the responsibility for the decline.

In Europe, there is clear evidence that many migrants have declined (Berthold *et al.* 1986). Whether this is due to a decrease in insect food in Africa due to a loss of trees and shrubs or whether it is due to a widening of the Sahara so that birds are unable to cross it is unknown (Lövei 1989). Palearctic migrants rarely use rainforests in Africa. Wood warblers *Phylloscopis sibilatrix* and honey buzzards *Pernis apivorus* occur there but also readily use a range of other habitats. In fact many migrants may even benefit from deforestation, as they use the agricultural habitats that replace the forest.

Many of the declines in European migrants have been linked to rainfall patterns in Africa and it is assumed that more suitable habitat is available in years of higher rainfall. In 1968 whitethroats *Sylvia communis* were a common summer visitor in Europe but by 1969 the population had dropped by about 77% in Britain and has remained at this low level since. The decline was linked to droughts in the Sahel region (Winstanley *et al.* 1974). Between the 1983 and 1984 breeding seasons British sand martins *Riparia riparia* declined in both population size and body size (Jones 1987) which was also linked to the severe drought in the Sahel. Studies of declining populations of swallows *Hirundo rustica* and sedge warblers *Acrocephalus schoenobaenus* show that both the population size and the survival rate of adults is strongly related to the rainfall in their respective winter quarters in south and west Africa (Møller 1989; Peach *et al.* 1991).

There continues to be debate about whether there is sufficient evidence to link the loss of temperate estuaries, the loss of rain forests, or the droughts in the tropics to declines in numbers of temperate birds. Estuaries are one of the few wildernesses left in Europe, rainforests are of prime conservation importance, and the consequences of droughts and overgrazing in Africa are horrific from many perspectives. These arguments alone should be sufficient to prevent the destruction of these habitats regardless of whether they affect populations elsewhere.

11.6 The future

It is likely that migration patterns have been changing for thousands of years as a consequence of human activity. For example, the habitats surrounding the Mediterranean were severely degraded 2000 years ago. With further human-induced changes likely the future of many species may depend upon their ability to change.

The need for such adaptation is likely to increase as a result of global warming. Berthold (1993) speculates as to the likely consequences of global warming for European birds. The species in Europe that are likely to benefit are the resident species whose populations are depressed by severe winters. It is also likely that populations of partial migrants will be composed of a greater proportion of sedentary individuals and that short and medium distance migrants should be selected to migrate less far. After the recent mild winters there have been numerous reports in central Europe of short distance migrants wintering within their breeding areas. Blackbirds *Turdus merula* (Merkel and Merkel 1983) and starlings *Sturnus vulgaris* (Schwabl 1983) have also changed their migratory behaviour in response to man-made environmental changes in Europe. The success of the late-arriving, long-distance migrants depends upon the degree of competition from resident birds and short-distance migrants that have already established territories. For example, the blackcap is a known competitor of the garden warbler, which is a long-distance migrant (Berthold and Terrill 1991). Global warming is thus likely to be deleterious for long-distance migrants.

11.7 Summary

Numerous studies have shown that habitat loss results in local population declines. The consequences of habitat loss can be estimated both in terms of the populations within sites and for migratory populations. The consequences of habitat loss may be much wider than simply affecting the populations using the site at time of loss. Coping with habitat loss may require changes in migration routes for some species. Genetic models show that this change may often take a large number of generations, and under some circumstances the population is unlikely to evolve to adopt the new route.

12

Predation and human disturbance

12.1 Introduction

Consumers may themselves be attacked and eaten by predators. Not only may this reduce the numbers of consumers but it may also cause them to change their choice of patches which may have consequences for the size of a population that could be sustained in a site. In this chapter I will describe the evidence for a trade-off between resource abundance and predation risk and will suggest a way of considering the consequences of this for the population that can be sustained in a site.

The disturbance caused by humans to wildlife is currently a subject of concern and debate. Human disturbance is best considered as an example of predation risk and the same conceptual framework can be used for considering the consequences for animal populations. There are a number of uses of the term disturbance, for example, botanists often use it to describe the creation of bare ground suitable for colonization and thus disturbance is often beneficial to the conservation interest of sites. I will use the term disturbance to refer to adverse behavioural changes in animals as a result of predators or humans. Such disturbance is usually detrimental to the conservation interest.

12.2 Predation

Choice of feeding patches will not only depend upon food availability but also upon the degree of predation risk (Lima and Dill 1990; McNamara and Houston 1987, 1990). A number of studies, especially of fish, have shown that animals do distribute themselves in relation to both food and predation risk. When given a choice of compartments, minnows *Rhinichthys atratulus* selected both those compartments with extra prey and those without the predatory fish *Semotilus atromaculatus* (Ceri and Fraser 1983). The minnows were more likely to visit a compartment with a predator if it also contained extra prey. Young black surfperch *Embiotoca jacksoni* distribute themselves in relation to both prey density and the risk of predation from kelp bass *Paralabrax clathratus*

(Holbrook and Schmitt 1988). Juvenile creek chubs *Semotilus atromaculatus* given freedom to select patches differing in both the densities of resources (*Tubifex* worms) and predatory adult creek chubs showed that patches were selected that minimized the mortality risk per unit energy gain (Giliam and Fraser 1987). Bluegill sunfish *Lepomis macrochirus* move from the littoral zone vegetation to the pelagic zone when a few years old. The pelagic zone results in a higher intake rate and a higher growth rate but the risk of predation by largemouth bass *Micropterus salmoides* was 40–80 times higher in the open water. The age at which the shift takes place varies between 2 and 4 years between different lakes and is correlated with the density of largemouth bass (Werner *et al.* 1983; Werner and Hall 1988). Similar results were shown for arctic charr *Salvelinus alpinus* (L'Abée-Lund *et al.* 1993).

A trade-off between predation risk and feeding preferences has also been shown for other taxa. As examples, willow tits *Parus montanus* usually feed on the outside of trees, but in a year when pygmy owls *Glaucidium passerinum* had few small mammals to feed on and thus hunted birds, the tits shifted to the centre of the trees (Suhonen 1993). Larval tiger salamanders *Ambystoma tigrinum* prefer to feed in vegetated shallows but in the presence of predatory beetles they shift to deeper water (Holomuzki 1986). Caribou *Rangifer tarandus* may select feeding locations on the basis of both food availability and the risk of predation from wolves (Ferguson *et al.* 1988).

One way of quantifying the effects of this trade-off on populations is to consider the amount of undepleted resources in relation to the predation risk. The proportion of prey eaten by sticklebacks increased with the distance from a predator (Fig. 12.1) (Milinski 1985). The distribution of herbivorous minnows *Campostoma anomalum* is influenced by the distribution of the predatory bass

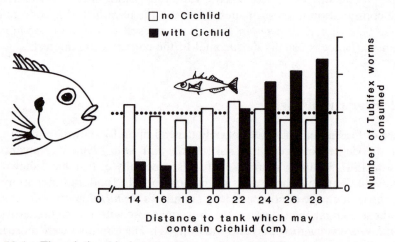

Fig. 12.1 The relationship between the number of *Tubifex* worms eaten by stickle-backs and the distance to a tank which sometimes contains a cichlid predator. (After Milinski 1985.)

Micropterus salmoides and *M. punculatus* (Power *et al.* 1985). The minnows deplete the algae population but are scarcer in areas which contain predators and hence the distribution of predators is reflected in the distribution of algae. White-throated sparrows *Zonotrichia albicollis* prefer to feed closer to shelter from predators and this is reflected in the pattern of seed depletion (Schneider 1984). Hence in each of these cases predation risk is reflected in the resource distribution. Thus, the total amount of resources available to the consumer may be reduced as some patches are underutilized as a result of predator avoidance. This will be the basis of the model developed in this chapter.

Abrahams and Dill (1989) quantified the trade-off between predation risk and intake rate for groups of ten guppies given a choice of two food patches, one of which contained a predatory fish. The predation risk was varied by moving the feeder so the guppies had to swim different distances to obtain the food. As expected, with higher predation risk, fewer individuals used the risky patch but those that did obtained a higher intake. This was then used to quantify how much extra food must be added to the risky patch in order that the guppies would use each patch equally. Adding extra food to the patch with the predator did offset the predation risk.

Individuals may differ in the manner by which they trade-off between food and predation risk. Abrahams and Dill (1989) showed such differences between male and female guppies; they suggested that the females were prepared to take a greater risk of predation because they benefited more from increased intake. A number of studies have shown that individuals are more likely to risk predation when their intake is low (Dill and Fraser 1984; Elgar 1986). McNamara and Houston (1990) provide state-dependent ideal free models incorporating predation risk; the choice of patch in these models depends upon the reward rate in each patch, the predation risk in each patch, and the individual's internal state. When individuals are hungry they are more likely to feed in the patch with the higher predation risk. Under such conditions intake rate, predation risk, and total mortality are expected to differ between patches, contrary to the simple expectations of the ideal free distribution.

12.3 Human disturbance

With human populations, recreation, and interest in conservation all increasing, there is mounting concern over the problem of human disturbance to wildlife (for example, Davidson and Rothwell 1993). There is good evidence that disturbance may influence breeding success [reviewed by Hockin *et al.* (1992)]. It is conceptually straightforward to consider a loss of breeding output due to disturbance and then to examine the consequences for population size, although in practice such measures may be difficult. However, in this chapter I will only consider the consequences of disturbance in relation to the ability to find food.

Disturbance may cause a species to avoid completely patches that they would otherwise use. This is conceptually straightforward as disturbance is equivalent to habitat loss, and the approaches described in the previous chapter can then be used. However, it is probably more common for disturbance to result in some patches being used less than expected. Human disturbance can be considered as equivalent to predation risk with consumers underutilizing disturbed patches.

It is very clear that disturbance can result in a redistribution. As a result of an increase in para-gliding in the Swiss Alps, chamois *Rupicapra rupicapra* spend more time in the forest (Schnidrig-Petrig *et al.* 1993). Creating experimental refuges from shooting in Britain and Denmark greatly increased the number of wildfowl in these areas (Hirons and Thomas 1993; Madsen 1993). When public access to a foreshore was prevented during land claim operations, the numbers of bar-tailed godwits *Limosa lapponica* feeding there increased by 50% (Furness 1973). The fact that redistribution is occurring may cause concern if the aim is to maintain the consumers within a particular area; for example, disturbance in a nature reserve may reduce the number of animals visible to the public or it may cause consumers to use areas where they are subject to hunting or pollution. However, redistribution does not necessarily mean that there are any adverse consequences for the total population size.

12.4 Theory of predation and disturbance

Disturbance is often assessed by determining either the distance from the disturbance at which individuals move away or the time before they return after being disturbed. Such studies are useful for comparing either the susceptibility of different species to disturbance or for comparing different sources of disturbance. For example, one admirable study showed that jogging and grass mowing disturbed coastal birds while bird watching did not (Burger 1981). However, the observation that individuals move from a source of disturbance and do not return for hours or even days is not necessarily a conservation problem if they do return at some point later and exploit the food. The precise order in which patches are exploited may often be irrelevant. If however interference is high, or the animals are territorial and the population is normally well dispersed between patches, then disturbance may result in individuals feeding at a higher local density than they would do otherwise and this may result in higher levels of starvation. Similarly, disturbance can be a problem if it encourages individuals to make greater use of patches with a higher risk of predation.

I will consider the consequences of predation or disturbance by extending the Sutherland and Anderson (1993) model of depletion outlined in Chapter 3. In that model, individuals feed in patches within a site until the resource in each patch reaches a certain threshold d_c, at which the intake would be insufficient for them to stay alive. As shown earlier, numerous studies have shown how consumers select patches by balancing the benefits of resource abundance and predation avoidance. Thus, at highly disturbed patches the threshold resource density is likely to be greater than in undisturbed patches. The consequences of this are illustrated in Fig. 12.2. If the species is insensitive to disturbance then all patches can be depleted to a low level and thus many consumers can survive in the site. If the species is sensitive to disturbance then resources may remain unexploited in the disturbed sites and fewer consumers can survive in the site. Obviously the greater the disturbance and the greater the susceptibility to disturbance, the fewer the consumers that can be sustained in a site. Another reason why the threshold resource density may be linked to disturbance is that the increased energy expenditure due to disturbance (Bélanger and Bédard 1989; White-Robinson 1982) may increase the threshold intake necessary for survival.

The empirical evidence is that d_c, the threshold to which consumers will feed, will be modified by disturbance. The form of this relationship is unclear so I have taken the simplest case and assumed that the threshold at which resources are abandoned equals $d_c (1 + \beta D)$, but of course, any empirical relationship could be added here. In this formulation, D is a measure of the frequency or severity of the disturbance or predation risk. This might be measured as the

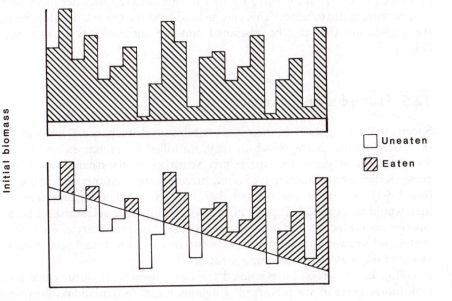

Fig. 12.2 The relationship between disturbance and the resource threshold at which a patch is abandoned. This shows an example of a series of patches at different distances away from a source of disturbance. The patches vary in the biomass of the resource they contain. This shows two different sensitivities to disturbance. In the upper figure disturbance has no impact on the threshold level at which the patch is no longer used. In the lower figure highly disturbed patches are abandoned at a high threshold level. The shaded area shows the resources that can be exploited.

number of predators, the distance to human disturbance, or the number of humans walking, driving, or shooting nearby. β is the susceptibility to disturbance, i.e. the extent to which disturbance raises the threshold prey density at which consumers stop feeding. This measure will vary greatly according to the species being studied and the type of disturbance. This can then be used to show the link between disturbance and the population that can be sustained in a site.

Consider a site in which each patch has an initial resource density j and a degree of disturbance D and so the frequency of each patch combination is $f_{j,D}$. Thus the amount of food eaten in each patch is a function of the initial resource abundance and the degree of disturbance. Then the maximum number of consumers, P that can be sustained in the site is:

$$P < \frac{\sum_{D=0}^{\infty}\left[T_h \sum_{d_c(1+\beta D)}^{M} \left\{ j - d_c(1+\beta D) \right\} f_{j,D} + 1/a' \sum_{d_c(1+\beta D)}^{M} f_{j,D} \log j/d_c(1+\beta D) \right]}{S} \tag{12.1}$$

This equation is simply eqn 3.6 with the threshold resource density modified to incorporate disturbance. This can be used to show the relationship between the population that can be sustained and the susceptibility to disturbance (Fig. 12.3).

12.5 Examples of disturbance

Species will vary greatly in their susceptibility to disturbance. This is likely to be related to the extent to which they are killed by predators and humans. Some species of geese are particularly sensitive to disturbance. Pink-footed geese (Keller 1990; Madsen 1985), red-breasted geese (Sutherland and Crockford 1993), and brent geese (Stock 1993) spend less time feeding near roads than would be expected. Tuite *et al.* (1984) examined the relationships between environmental factors and numbers of waterfowl on reservoirs in England and Wales and showed that certain susceptible species were found less often than expected on sites with intensive recreation.

Gill *et al.* (in press) have studied the consequences of disturbance on the pink-footed geese in north Norfolk, England, which feed inland on the remains of harvested sugar beet *Beta vulgaris*. These geese may be particularly sensitive to disturbance as they are shot when feeding on cereal and when they fly to and from the offshore roost. When foraging on sugar beet fields, geese were disturbed significantly more often where they fed closer to roads.

By measuring the biomass of sugar beet fragments before and after the geese fed, it was discovered that a higher percentage of the sugar beet was depleted in those fields where the geese were further away from the road (Fig. 12.4). The total number of bird-days depended on both the distance from the road

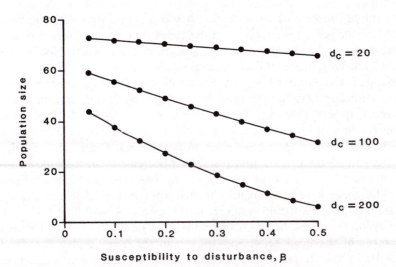

Fig. 12.3 The relationship between the susceptibility to disturbance and the consumer population that can be sustained. This uses eqn 12.1 to calculate the population that can be sustained in a site. The model assumes there are 100 patches, each of 1000 m². There are ten levels of resource density, j, varying linearly between 100 and 1000 items per m². There are also ten different disturbance levels, D, varying linearly between 1 and 10. Each patch has a different combination of resource density and disturbance level. The handling time is 5 s and the attack constant a' is 0.0001. The threshold resource density for survival, d_c, varies between 20 and 200 items m⁻² in the different simulations. The season, S, is 100 days of 10 hours foraging per day.

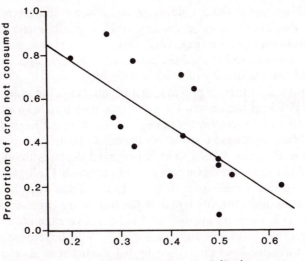

Fig. 12.4 The percentage of sugar beet biomass depleted by pink-footed geese in relation to the distance from the road. Each point is a separate field. (From Gill *et al.*, in press)

and the initial biomass of sugar beet. They calculated that if all fields were exploited by the geese to the level of those fields that were disturbed least then the total number of geese that could be sustained in the North Norfolk study area would be 64% higher (Gill *et al.*, in press).

A change in the nature of the disturbance may well be expressed in the slope of the relationship between the amount of sugar beet eaten and the distance to the road. If, say, there is a large increase in the number of birdwatchers then this may result in an increased slope as the geese avoid fields adjacent to the road. Should the slope of this relationship increase from the −1.51 shown in Fig. 12.4 to −3 then this would result in a decrease of 61% in the numbers of geese that could be sustained. Similarly if the there was a reduction in hunting so that the geese were less sensitive to disturbance, and if this changed the slope to −0.6 then this would result in an increase of 42% (Gill 1994).

As a second example of how disturbance can be modelled I will return to the example of the bean geese and wigeon in the Yare Valley described in Chapter 10. In this chapter I extend this model to include the consequences of disturbance on the wigeon population. In an attempt to quantify disturbance, two wardens and Gary Allport independently gave each field a relative index of disturbance. This technique is obviously extremely simplistic but has the great merit of attempting to combine sources of disturbance such as a birdwatching hide, a railway line, farm tracks and a river used by holiday-makers.

The distribution of the birds will also depend upon the amount of grass in each field. Wigeon have a clear preference for areas with an intermediate biomass of about 100 g dry weight per m^2; they tend to ignore areas of much longer or shorter grass (see Chapter 10). The problem is then to find a common currency which combines the avoidance of disturbance and the preference for a certain biomass. The preferences for areas differing in biomass are expressed in terms of droppings per m^2 (Fig. 10.2) and I assumed that the disturbance could simply be subtracted from these preferences. By this approach the units for disturbance are the density of wigeon droppings!

A major problem is then converting disturbance into the same units as the preference for different biomasses. Thus one field may have a sward of a height that gives a value of 15 wigeon droppings m^2 but have a disturbance level of 100. A second field may have a sward equivalent to 10 droppings m^{-2} but have a disturbance level of 20. Which field is preferred depends upon the relative importance of biomass and disturbance. The approach I adopted was to use a range of conversion values from 0.000 01 to 0.1. Thus with the disturbance index in each field multiplied by 0.000 01 the role of disturbance is trivial and biomass is the sole determining factor. In the above example the first field would have a value of 14.999 and the second would have a value of 9.998. With each disturbance level multiplied by 0.1 disturbance is overwhelmingly important. The first field will have a value of 5 and the second a value of 8.

What then is the correct conversion value? The model was run with a range of disturbance conversion values and the relationship between the observed

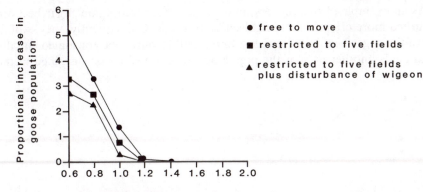

Fig. 12.5 The relationship between the number of bean geese that can be sustained and the increase in the wigeon population as described in Chapter 10. The three curves represent the following scenarios: ● both wigeon and bean geese choose fields solely on the basis of sward biomass; ■ wigeon select fields solely on the basis of sward biomass and bean geese are restricted by disturbance to the five least-disturbed fields; ▲ wigeon select fields according to a trade-off between disturbance and sward biomass and bean geese are again restricted to the five least-disturbed fields.

and predicted distributions of wigeon was determined. The lowest conversion values in which biomass is of predominant importance gave a Spearman rank correlation coefficient of 0.39 while the highest values in which disturbance is particularly important gave a coefficient of 0.40. Intermediate values around 0.005 gave a correlation of 0.63. This suggested that incorporating disturbance significantly improved our description of the system and that even simple means of combining disturbance and biomass may be useful.

It is then possible to assess the consequences of this disturbance for the bean goose population. As Fig. 12.5 shows, incorporating disturbance reduces the goose population that can be sustained. The reason for this is that disturbance results in wigeon being more likely to use the undisturbed fields in the centre of the study site which are preferred by the bean geese. The wigeon then deplete the vegetation, making it unsuitable for the geese.

12.6 Summary

Human disturbance can be considered as a form of predation risk. Both human disturbance and predation can result in patches being underused. This can be quantified by relating the threshold resource density at which consumers cease feeding to the degree of disturbance. This can then be used to determine the population that may be sustained in a site in relation to the density of resources and the degree of disturbance.

As an example of this process, pink-footed geese feeding on sugar beet are disturbed more often in those fields close to roads. Consequently, they exploit less of the sugar beet in these fields. Should all resources be consumed equally in all fields another 64% of the population of geese could be sustained in the area.

13

Modelling techniques

13.1 Introduction

The aim of this chapter is to outline the basic approaches behind the models described in the previous chapters. There is not the space to describe every model in detail but my hope is that this chapter will make it reasonably clear how these models were created and assist anyone wanting to produce further models. The models used in this book were all programmed in Pascal and are not complex.

All the models have a similar basic structure. In each case there are patches that differ in resource density and consumers distribute themselves between these patches in response to the gains available from each. These models are based on the ideal free distribution and so the assumption is that no individual can obtain higher rewards by moving elsewhere. Thus the evolutionarily stable strategy is that, for each phenotype, the gains are equal in all the patches occupied. If the gain is higher in one patch then the theoretical expectation is that individuals will move there until either the increased negative feedback from interference and depletion balances the discrepancy or all members of that phenotype have moved.

13.2 Interference models

Interference is the reduction in the ability to find resources due to the presence of others. It may be considered as a reduction in the attack constant a' which is the product of the area or volume of substrate searched in time T and the probability of detecting and obtaining those resource items within this area or volume. Equation 2.1 gives one formulation for describing the reduction in the attack constant as a result of interference. If all individuals are equal then the distribution of consumers between patches is given by eqn 2.2.

Individuals will differ in the interference they experience, for example some individuals will be dominant over others. The extent of interference experienced by a given individual is likely to depend upon the average level of interference, the number of competitors and its competitive ability relative to that of others. Equation 2.3 can be used to describe the intake obtained by a given individual but other formulations and approaches should be explored.

The models of lekking described in Chapter 7 also use the approach described here.

1. Consider a site consisting of a number of patches differing in resource density.

2. For incorporating variation in competitive ability consider a range of values of competitive ability (e.g. 0.8 to 1.2). It is easiest to start with a small number of phenotype classes (i.e. fewer than 10). Decide upon the number of individuals allocated to each class, for example this may follow a normal distribution of the number of individuals in each class.

3. Assume an initial distribution such that the same proportion of each phenotype is located in each patch (use an array to state numbers of each phenotype in each patch). For example, with 5 phenotypes each of 200 individuals and 10 patches then there are initially 20 individuals of each phenotype in each patch.

4. Determine for each phenotype the intake in each patch and create a second array for these values. In order to calculate intake, for each patch determine the density and mean competitive ability and thus the interference experienced by each individual. From the value of a' and the resource density, the intake rate can be calculated. One approach is to assume that the number of resource items eaten Na in time T is related to the resource density by the following equation:

$$\frac{Na}{T} = \frac{a'\alpha}{1 + a'\alpha\, H_t} \tag{13.1}$$

where H_t is the handling time—the time between finding one food item and commencing searching for the next (Holling 1959). The effect of a decline in the attack constant due to interference on intake rate can then be determined.

5. The evolutionarily stable strategy is determined by assuming that individuals move in relation to the gains in each patch. To achieve this multiply the number of individuals of each phenotype in each patch by the intake of that phenotype in the patch (i.e. create a third array by multiplying together the arrays described in steps 3 and 4). Thus if there are 20 individuals of a phenotype in a patch and their intake is 2.36 then the new number of individuals is 47.2.

6. Calculate the new total number of each phenotype in all the patches by summing the values calculated in step 5 across all patches. In this example, the number of one phenotype in one patch is 47.2 and let us assume the total of that phenotype in ten patches is 411.8. The numbers of each

phenotype in each patch need to be adjusted to the total number that are actually present. This is achieved by dividing the actual total (i.e. 200) by the new calculated total (i.e. 411.8). Thus in this case 200 ÷ 411.8 gives 0.485. The numbers of the phenotype in each patch are then multiplied by this figure (in this case $47.2 \times 0.485 = 22.89$). The total number of the phenotype across all patches is now restored to 200 but the distribution has changed so that for each phenotype more individuals are in the patches where their intake was highest. This procedure has to be repeated for each phenotype.

7. Steps 4–6 are repeated until the distribution is stable. It is useful to print out the arrays described in steps 3 and 4 to show when stability occurs. Steps 5–7 ensure that the intake is the same for all members of a phenotype. At this point the intake for a particular phenotype is equal in all the sites in which individuals of that phenotype are present.

13.3 Depletion models

Under simple cases the equations given in Section 3.2 describe the pattern of resource depletion and the number of individuals a site can sustain. Simulation models are used for more complex cases and are described below.

1. Consider a site divided into patches each of a given size and given initial resource density. Patches with the same initial resource density can be combined as a larger patch.

2. It is necessary to consider what determines intake. One approach is to assume that Holling's (1959) disk equation (see Section 13.2) provides a good description of the intake.

   A common mistake is to use values of a' that are much too high. a' multiplied by the resource density gives the number of items found per second spent searching, and thus a very low value such as 0.000 01 is often realistic. If, as is usually the case, a' is expressed in area searched per second then the time T needs also to be in seconds.

   An alternative to Holling's disc equation is to quantify what determines intake and build this into the model. A third approach is to assume that, although consumers feed in the sites with the most resources the level of depletion is relatively constant (as for many vertebrates the variation in daily intake may be relatively small). This technique was used in the models of bean geese given in Chapter 10.

3. Assuming that there is no interference (see Section 13.4) and that resource density is the major factor determining the consumer distribution the approach is then to assume all individuals feed in the patch with the highest resource density. For any given system other factors will be important and if these can be quantified they can be incorporated at this stage (Section

13.9). The depletion caused in the time interval is then calculated and this is subtracted from the resource density.

4. Repeat steps 2 and 3 for each time interval. As depletion proceeds the resource density becomes equal across the patches. The assumption that the consumers always feed in the patch with the highest resource density may seem reasonable for gregarious species but not for solitary species, who might be expected to be distributed between patches. With this modelling approach the assumption is that the entire population of consumers will shift continually between patches. In practice these both give the same answer of the same mean number of individuals per patch. If the time interval used is too long excess localized depletion can take place.

13.4 Combining interference and depletion

This simply combines the techniques of the two approaches described above. As the evolutionarily stable strategy has to be repeated for each time interval, models combining interference and depletion can be time-consuming if there are many phenotype classes, patches, and time intervals.

1. Consider a site with patches each of known size differing in resource density.

2. Divide the total time into intervals.

3. Determine for one time interval the number of individuals in each patch and their intake using the approach described in Section 13.2.

4. Determine the depletion in each patch and subtract this from the resource density.

5. Repeat steps 3 and 4 for each time interval.

13.5 Prey availability

1. Consider a site comprising a series of patches that may differ in resource density.

2. Within each patch some prey may be easier to capture than others and this may be expressed as differences in the attack constant a'. The number of prey α_i in each availability class, i, in each patch is given in an array.

3. The intake rate can be calculated by modifying Holling's disc equation (see Section 13.2) so that the attack rate of each availablity class is a'_i.

$$\frac{Na}{T} = \frac{\sum\limits_{i=1}^{n} a'_i \, \alpha_i}{1 + \sum\limits_{i=1}^{n} a'_i \, \alpha_i \, H_t} \, . \tag{13.2}$$

4. Assume individuals feed in the patch with the highest intake (as in Section 13.3) and calculate the resulting depletion of each availability class.

5. Steps 3 to 4 are repeated for all time intervals.

13.6 Density-dependent mortality

1. Create a model incorporating the relevant components given in Sections 13.2 to 13.5.

2. Incorporate a threshold level of intake and assume that if individuals of a phenotype class have an intake below this threshold then some or all will die. If some individuals die then the intake of others may increase due to reduced interference and this may bring them back above the threshold intake. If however the model has only a few phenotype classes, and entire classes die at a time, the mortality will occur in steps. To remove this problem it is necessary to either have numerous phenotype classes or for only a fraction of each class to die at a time. In both cases once mortality is inflicted the distribution and intakes are recalculated. Repeat this until all remaining individuals stay alive. After this calculate the depletion caused by the surviving individuals in the time interval.

3. Run the simulation for a range of initial consumer densities to determine the relationship between density and mortality.

13.7 Equilibrium population size

The equilibrium population size is the population size at which the birth rate and death rate are equal, this will depend upon the degree of density dependence (see Fig. 1.1). The equilibrium population size can be determined graphically (Fig. 6.14) or by creating a difference equation model relating the number of individuals N at year $t + 1$ to the number in the previous year t by

$$N_{t+1} = N_t + B - D \tag{13.3}$$

where B is the number of births and D the number of deaths. The relationship between birth rate and density can be assessed from the extent of territorial behaviour (see Chapter 7), and the relationship between death rate and density assessed from the nature of interference and depletion (see Chapter 8). Such subcomponents are best created independently before being combined. This model can be readily extended to consider different ages and sexes.

Starting from an initial density the equation is repeatedly run until an equilibrium is reached. However, if the mortality or birth rate is strongly overcompensating then cycles or chaos may result (May 1974).

13.8 Migration routes

1. Consider a range of breeding sites and wintering sites. The nature of density-dependent mortality and breeding output in each must be determined. One possibility is to incorporate the techniques given in Sections 13.2, 13.4, and 13.6.

2. Consider the costs of moving between each breeding and wintering site. This may be a mortality cost, or as Sutherland and Dolman (1994) used, an energy cost which requires an increase in threshold intake in the wintering site.

3. Produce an array giving the number of individuals moving between each pair of breeding and wintering sites. Start with an equal number adopting each route.

4. Determine the number breeding in each site and the resultant birth rate. These are distributed between wintering sites according to the gain from each in the previous year. In the first year they are distributed equally.

5. Determine the number wintering in each site and the resultant mortality.

6. Repeat steps 4 and 5 until an equilibrium is reached.

13.9 Models of specific systems

The techniques described above give the general techniques. For a model of a particular species in a particular area it is necessary to incorporate the details of ecology. As an example of the approach Fig. 10.1 gives the flow diagram used for modelling the distribution of bean geese and wigeon.

1. Consider what determines the distribution. As well as resource density there may also be resource quality, tidal cover, or disturbance. These relationships may be complex, for example, in the study at Buckenham, bean geese prefer the longest sward available while wigeon prefer an intermediate length sward.

2. For each patch quantify those features shown to be important in step 1, e.g. biomass, time covered by tide, or disturbance.

3. For each time interval determine where individuals feed according to the preferences quantified in step 1. If the area is tidal it is necessary to determine the distribution for different stages of the tide during which different combinations of patches are exposed.

4. Calculate the depletion in each patch.

5. Incorporate any growth or reproduction of the vegetation or prey populations. This may vary between patches or with the season.

6. Repeat steps 3–5 for each time interval.

14

Summary

14.1 Introduction

The aim of this book has been to show how simple models based on the decisions individuals make can be used to describe phenomena such as the distribution within and between sites, population abundance, and migration. It is then possible to use these models to explore conservation issues such as the consequences of habitat loss or changes in habitat management.

Population ecology has usually been based upon empirical relationships. This has the disadvantage that if the environment changes it is necessary to redetermine these relationships. Behavioural models based upon optimization and game theory can be derived from the principles of natural selection. Deriving population models from behaviour thus gives the opportunity to consider population ecology from first principles.

The central idea behind this book is that of the ideal free distribution. This simple concept considers the distribution of animals in relation to resource density and the distribution of other individuals. The expectation of the ideal free distribution is that individuals will distribute themselves so that the rewards are equal in all the patches they occupy. Obviously if the rewards were greater in one patch then individuals could benefit by moving there until the negative feedback, such as interference or depletion, balances the rewards.

The ideal free distribution provides a good starting point, but in reality many aspects of natural history mean that it is unrealistic. Much of this book shows how the concept of the ideal free distribution can be made more realistic.

The link between behaviour and population ecology can be divided into two steps. Firstly models of behaviour can be used to predict density dependence. Models of foraging behaviour can be used to describe density-dependent mortality as at high populations there will be reduced intake due to enhanced interference, depletion, and use of poorer sites (the buffer effect), and by making the reasonable assumption that individuals die if they fail to obtain a sufficient intake, it is possible to examine how the mortality changes with density. Models of breeding behaviour also demonstrate the link with density-dependent fecundity as at high populations, individuals occupy poorer patches, may refrain from breeding, or occupy smaller territories. The second step extends these density-dependent relationships to models of population sizes, migration routes, and the response of populations to habitat loss. Thus it is possible to relate population size to aspects of behaviour.

14.2 Decision making

There is a great potential for studies examining the decisions made by identifiable individuals. This would be particularly interesting if carried out in the field and combined with experiments. Future work could examine in more detail the costs and benefits for given individuals of feeding in different patches. During an experiment each will experience different intakes partly as a result of chance. It would be interesting to study which individuals switch patches and when. One easy approach in studies of continuous input, such as those with fish or ducks, would be to feed specific individuals with different schedules to determine the conditions under which individuals move.

14.3 General and specific models

In this book I describe two types of model. One approach is to produce general models of processes such as territoriality, population regulation, and migration to show the general relationships between behaviour and populations. Of course, for any particular species there will be some details of the ecology that mean that such a general model is not strictly applicable. The second approach is to provide models of specific systems. Although natural history is currently unfashionable amongst biologists and is often used as a pejorative term, in my experience the details of the natural history are essential for applying ecological theory. For example, understanding how to conserve bean geese and wigeon in England (Chapter 10) required the knowledge that bean geese selected the taller swards while wigeon selected intermediate swards. Understanding the response of brent geese to their own increasing population required understanding of their preferences for different intertidal habitats, the nature of productivity of these plant communities, and the loss of algae through storms. Comprehension of the breeding performance of oystercatchers required understanding of the distinction between 'residents' (breeding next to the mud-flats) and 'leap-frogs' (breeding inland). In each case generalized models provide the framework, but a specific model incorporating the detail is essential to understand the system.

14.4 Density dependence

Interference is the decline in intake resulting from the behaviour of others. By measuring the strength of interference it is possible to predict the expected distribution. However, there are remarkably few field measures of the strength of interference in vertebrates. There is also a need for detailed studies of the mechanisms of interference and further theory to describe exactly how interference affects different individuals.

One recurrent theme throughout this book is the importance of individual variation within both the resource and consumer populations. Individual differences between consumers in the ability to find food, the susceptibility to interference, and tollerance of low intake all have considerable consequences for the nature of density-dependent mortality. There is considerable potential here for field studies combining physiological and ecological techniques. The origin of this variation is also poorly understood. To what extent is it genetic or environmental? Are different individuals adopting different strategies? How important is parasite load? More studies linking the feeding biology outside the breeding season to the reproductive success would be useful here.

Depletion, the removal of resources, obviously occurs in all situations in which consumers are feeding on resources. It is theoretically less elegant than interference but may well be of much greater importance.

In many species much of the density-dependent mortality may occur in short periods of severe weather or food shortage. In a research programme lasting a few years such an event is likely to be missed or to be considered an atypical disaster.

It is convenient to assume that all prey are equally available. However, numerous studies have shown that some prey are more accessible. For example, some may be nearer the surface of the substrate. This causes considerable problems for the interpretation of field data. The biologist's measure of resource density may be very different from the animal's measure of available resources. I believe the importance of this problem is greatly underrated.

The distribution of consumers has obvious implications for the spatial pattern of prey mortality. Whether the mortality is density dependent, inversely density dependent, or density independent depends upon the values of aspects of behaviour such as the level of interference or the degree of variation between individuals.

14.5 Territoriality

A different approach is required to consider territoriality. I show how an understanding of the costs and benefits of territory size can be used to examine the nature of density-dependent breeding output. At higher densities, individuals are more likely to occupy poorer patches, have smaller territories, or defer breeding to another year, all of which result in a reduced breeding output. As an example of this approach, I produce a preliminary model to show how the sequence in which oystercatchers occupy different quality territories leads to density-dependent breeding output. This can be combined with estimates of winter mortality to show the expected equilibrium population size.

From such a model it is possible to explore the consequences for the population of an increase in mortality, as for example may occur due to increased hunting or as a result of increased competition for food as a consequence of habitat loss.

14.6 Response to environmental change

Habitat is continually being lost due to built developments and changes in management practices. In Europe much of this can be attributed to the disasterous Common Agricultural Policy. Global climate change and sea level rise may result in further changes. With the models described in this book it is possible to determine the equilibrium population size of populations and examine the consequence of losing part of their habitat.

It is particularly complex to consider the consequences of habitat loss on migratory populations. The approach used for considering the choice of patches within a site can be used on a larger scale for considering the distribution between patches. With a range of breeding sites and wintering sites it is possible to determine the evolutionarily stable migration strategies that link these sites. It is then possible to rerun such models with part or all of one or more sites missing and determine the consequences for the equilibrium population of habitat loss.

One assumption of this approach is that the equilibrium has been reached. It may be the case that the change in the habitat may occur at too fast a rate for some species to respond and extinction may occur first. There is evidence that some species have altered their migration route in the recent past in response to environmental changes. However some European migrant birds adopt very peculiar routes and one interpretation is that these routes are a consequence of the distribution in the last ice age.

Housing or industrial development may result in direct loss of habitat. In many cases the associated increase in disturbance is of equal concern. Many studies have shown that the distribution of individuals is best thought of not only in response to the resource distribution but also in relation to the distribution of predators. The same approach can be used to consider the response to human disturbance. More of the resource is left uneaten in the areas of high-predation risk or disturbance. It is thus possible to examine the consequence of disturbance for the population size that can be sustained.

14.7 Further applications

Most of the specific models described in this book involve birds as I am primarily an ornithologist, but there is no reason why similar models cannot be applied to very different groups. For example, it was relatively easy to use such an approach to model hyenas, and I am collaborating with a colleague, John Reynolds, who is a fish biologist, to model density dependence and population size in bitterling *Rhodeus sericeus* populations using the same approach as was used for the model of oystercatcher territories.

I have restricted this book to considering vertebrates. Clearly some of the phenomena described here apply equally well to invertebrates. For example some invertebrate species migrate, defend territories, or lek. Such an approach can also be extended to describe plants in which the vegetative growth via stolons or rhizomes can be considered equivalent to the search path of a foraging animal (Sutherland and Stillman 1988). Finally the approach used in this book, with appropriate modification, could also be used by anthropologists for considering the distribution and abundance of some human cultures.

References

Abrahams, M.V. (1986) Patch choice under perceptual constraints: a cause for deviations from an ideal free distribution. *Behavioural Ecology and Sociobiology*, **19**, 409–15.

Abrahams, M.V. (1989) Foraging guppies and the ideal free distribution—the influence of information on patch choice. *Ethology*, **82**, 116–26.

Abrahams, M.V. and Dill, L.M. (1989) A determination of the energetic equivalence of the risk of predation. *Ecology*, **70**, 999–1007.

Abrahams, M.V. and Healey, M.C. (1990) Variation in the competitive abilities of fishermen and its influence on the spatial distribution of the British Columbian salmon troll fleet. *Canadian Journal of Fisheries and Aquatic Sciences*, **6**, 1116–21.

Adriaensen, F. and Dhondt, A.A. (1990) Population dynamics and partial migration of the European robin (*Erithacus rubecula*) in different habitats. *Journal of Animal Ecology*, **59**, 1077–90.

Adriaensen, F., Ulenaers, P. and Dhondt, A.A. (1993) Ringing recoveries and the increase in numbers of European great crested grebes (*Podiceps cristatus*). *Ardea*, **81**, 59–70.

Alatalo, R.V., Glynn, C. and Lundberg, A. (1986) Female pied flycatchers choose territory quality not male characteristics. *Nature*, **323**, 152–3.

Alatalo, R.V., Lundberg, A., Höglund, J. and Sutherland, W.J. (1992) Evolution of Black grouse leks—female preferences benefit males in larger leks. *Behavioural Ecology*, **3**, 53–9.

Alatalo, R.V., Lundberg, A. and Ulfstrand, S. (1985) Habitat selection in the pied flycatcher *Ficedula hypoleuca*. In *Habitat selection in birds* (ed. M.L. Cody), (pp. 59–83). London: Academic Press.

Alerstam, T. (1990) *Bird Migration*. Cambridge: Cambridge University Press.

Alerstam, T. and Enckell, P.H. (1979) Unpredictable environments and evolution of bird migration. *Oikos*, **33**, 228–32.

Alerstam, T. and Högstedt, G. (1982) Bird migration and reproduction in relation to habitats for survival and breeding. *Ornis Scandinavica*, **13**, 25–37.

Alexander, R.D. (1974) The evolution of social behaviour. *Annual Review of Ecology and Systematics*, **5**, 325–83.

Alexander, R.D. (1975) Natural selection and specialised chorusing behaviour in acoustical insects. In Pimental (ed.), *Insects, science and society*. (pp. 35–77). New York: Academic Press.

Allee, W.C., Emerson, A.E., Park, O., Park, T. and Schmidt, K.P. (1949) *Principles of Animal Ecology*. Philadelphia: Saunders.

Allport, G.A. (1991) The feeding ecology and habitat requirements of overwintering western taiga bean geese (*Anser fabalis fabilis*). Ph.D. thesis, University of East Anglia.

Andrén, H. (1990) Despotic distribution, unequal reproductive success and population regulation in the jay *Garrulus glandarius* L. *Ecology*, **71**, 1796–1803.

Andrén, H. and Angelstam, P. (1988) Elevated predation rates as an edge effect in habitat islands: experimental evidence. *Ecology*, **69**, 544–7.

Arcese, P. and Smith, J.N.M. (1985) Phenotypic correlates and ecological consequences of dominance in song sparrows. *Journal of Animal Ecology*, **54**, 817–30.

Arditi, R. and Akçakaya, H.R. (1990) Underestimation of mutual interference by predators. *Oecologia*, **83**, 358–61.

Arditi, R., Ginzburg, L.R. and Perrin, N. (1992) Scale invariance is a reasonable approximation in predation models—reply to Ruxton and Gurney. *Oikos*, **65**, 336–7.

Arnold, G.W. (1964) Factors within plant associations affecting the behaviour and performance of grazing animals. In D.J. Crisp (ed.), *Grazing in terrestrial and marine environments* (pp. 133–54). Oxford: Blackwell Scientific.

Åström, M. (1994) Travel cost and the ideal free distribution. *Oikos*, **69**, 516–9.

Baeyens, G. (1981) Functional aspects of serial monogamy—the magpie pair-bond in relation to its territorial system. *Ardea*, **69**, 145–66.

Bailey, N.T.J. (1962) Interactions between hosts and parasites when some individuals are more difficult to find than others. *Journal of Theoretical Biology*, **3**, 1–18.

Baker, M.C., Belcher, C.S., Deutsch, L.C., Sherman, G.L. and Thompson, D.B. (1981) Foraging success in junco flocks and the effects of social hierarchy. *Animal Behaviour*, **29**, 137–42.

Baker, R.R. (1978) *The evolutionary ecology of animal migration*. London: Hodder and Stoughton.

Balmford, A., Albon, S. and Blakeham, S. (1992) Correlates of male mating success and female choice in a lek-breeding antelope. *Behavioural Ecology*, **3**, 112–23.

Barnard, C.J. (1984) The evolution of food-scrounging strategies within and between species. In C.J. Barnard (ed.), *Producers and scroungers* (pp. 95–126). London: Croom Helm.

Bateman, A.J. (1948) Intra-sexual selection in Drosphila. *Heredity*, **2**, 349–68.

Batten, L.A., Bibby, C.J., Clement, P., Elliott, G.D. and Porter, R.F. (1990) *Red data birds in Britain*. London: Poyser.

Beddington, J.R. (1975) Mutual interference between parasites or predators and its effect on searching efficiency. *Journal of Animal Ecology*, **44**, 331–40.

Beehler, B.M. and Foster, M.S. (1988) Hotshots, hotspots and female preferences in the organisation of lek mating systems. *American Naturalist*, **131**, 203–19.

Bélanger, L. and Bédard, J. (1989) Responses of staging greater snow geese to human disturbance. *Journal of Wildlife Management*, **53**, 713–19.

Bell, D.J. (1986) Social effects on physiology in the European Rabbit. *Mammal Review*, **16**, 131–7.

Bensch, S. and Hasselquist, D. (1991) Territory infidelity in the polygynous great reed warbler *Acrocephalus arundinaceus*: the effect of variation in territory attractiveness. *Journal of Animal Ecology*, **60**, 857–71.

Berg, Å., Lindberg, T. and Gunnar, K. (1992) Hatching success of lapwings on farmland: differences between habitats and colonies of different sizes. *Journal of Animal Ecology*, **61**, 469–76.

Bernstein, C., Kacelnik, A. and Krebs, J.R. (1988) Individual decisions and the distribution of predators in a patchy environment. *Journal of Animal Ecology*, **57**, 1007–26.

Bernstein, C., Kacelnik, A. and Krebs, J.R. (1991a) Individual decisions and the distribution of predators in a patchy environment. II The influence of travel costs and structure of the environment. *Journal of Animal Ecology*, **60**, 205–25.

Bernstein, C., Krebs, J.R. and Kacelnik, A. (1991b) Distribution of birds amongst habitats: theory and relevance to conservation. In C.M. Perrins, J.D. Lebreton, and G.J.M. Hirons (ed.), *Bird population studies* (pp. 317–45). Oxford: Oxford University Press.

Berthold, P. (1984) The control of partial migration in birds: a review. *Ring*, **10**, 253–65.

Berthold, P. (1988) Evolutionary aspects of migratory behaviour in European warblers. *Journal of Evolutionary Biology*, **1**, 195–209.

Berthold, P. (1993) *Bird migration*. Oxford: Oxford University Press.

Berthold, P. (1995) Microevolution of migratory behaviour illustrated by the blackcap *Sylvia atricapila*. *Bird Study* **42**, 89–100.

Berthold, P., Fliege, G., Querner, U. and Wimkler, H. (1986) Die Bestandsentwicklung von Kleinvögel in Mitteleuropa: analyse von fangzahlen. *Journal für Ornithologie*, **127**, 397–437.

Berthold, P., Helbig, A.J., Mohr, G. and Querner, U. (1992) Rapid microevolution of migratory behaviour in a wild bird species. *Nature*, **360**, 668–70.

Berthold, P., Mohr, G. and Querner, U. (1990) Steuerung und potentielle Evolutionsgeschwindigkeit des obligaten Teilzieherverhaltens: Ergebnisse eines Zweiweg—Selektionsexperiments mit de Mönchsgrasmücke (*Sylvia atricapilla*). *Journal für Ornithologie*, **131**, 33–45.

Berthold, P. and Terrill, S.B. (1991) Recent advances in studies of bird migration. *Annual Review of Ecology and Systematics*, **22**, 357–78.

Bertram, B.R.C. (1978) Living in groups: predators and prey. In J.R. Krebs and N.B. Davies (ed.), *Behavioural ecology* (pp. 64–96). Oxford: Blackwell Scientific.

Beukema, J.J., Essink, K., Michaelis, H. and Zwarts, L. (1993) Year-to-year variability in the biomass of macrobenthic animals on tidal flats of the Wadden Sea: how predictable is this food source for birds? *Netherlands Journal of Sea Research*, **31**, 319–30.

Bibby, C.J. and Green, R.E. (1980) Foraging behaviour of migrant pied flycatchers (*Fidedula hypoleuca*) on temporary territories. *Journal of Animal Ecology*, **49**, 507–21.

Bierbach, H. (1983) Genetic determination of partial migration in the European robin, *Erithaceus rubecula*. *Auk*, **100**, 601–6.

Bijlsma, R.G. (1990) Predation by large falcons on wintering waders on the Banc d'Arguin, Mauritania. *Ardea*, **78**, 75–82.

Black, J.M. and Owen, M. (1989) Agonistic behaviour in barnacle goose flocks: assessment, investment and reproductive success. *Animal Behaviour*, **37**, 199–209.

Black, R. (1971) Hatching success in the three-spined stickleback, *Gasterosteus aculeatus*, in relation to changes in behaviour during the parental phase. *Animal Behaviour*, **19**, 532–41.

Böhning-Gaese, K., Taper, K.L. and Brown, J.L. (1993) Are declines in North American insectivorous songbirds due to causes on the breeding range? *Conservation Biology*, **7**, 76–86.

Bradbury, J.W. (1977) Lek mating behaviour in the hammer headed bat. *Zeitschrift fur Tierpsychology*, **45**, 225–55.

Bradbury, J.W., Gibson, R.M., McCarthy, C.E. and Vehrencamp, S.L. (1989) Dispersion of displaying male sage grouse II The role of female dispersion. *Behavioural ecology and sociobiology*, **24**, 15–24.

Bradbury, J.W., Gibson, R.M. and Tsai, I.M. (1986) Hotspots and the evolution of leks. *Animal Behaviour*, **34**, 1694–709.

Brittingham, M.C. and Temple, S. (1988) Impacts of supplementary feeding on survival rates of black-capped chickadees. *Ecology*, **69**, 581–9.

Brockmann, H.J. and Barnard, C.J. (1979) Kleptoparasitism in birds. *Animal Behaviour*, **27**, 487–514.

Brooke, M.d.L. (1979) Differences in the quality of territories held by wheatears (*Oenanthe oenanthe*). *Journal of Animal Ecology*, **48**, 21–32.

Brown, J.L. (1964) The evolution of diversity in avian territorial systems. *Wilson Bulletin*, **76**, 160–9.

Brown, J.L. (1969*a*) The buffer effect and productivity in tit populations. *American Naturalist*, **103**, 347–54.

Brown, J.L. (1969*b*) Territorial behavior and population regulation in birds. *Wilson Bulletin*, **81**, 293–329.

Bryant, D.M. (1979) Effects of prey density and site character on estuary usage by overwintering waders (Charadrii). *Estuarine and Coastal Marine Science*, **9**, 369–84.

Bryant, D.M. (1987) Wading birds and wildfowl of the estuary and Firth of Forth, Scotland. *Proceedings of the Royal Society of Edinburgh*, **93B**, 509–20.

Burger, J. (1981) The effects of human activity on birds at a coastal bay. *Biological Conservation*, **21**, 231–41.

Burger, J. and Gochfield, M. (1983) Feeding behaviour in laughing gulls: compensatory site selection by young. *Condor*, **85**, 467–73.

Buxton, N.E. (1981) The importance of food in the determination of the winter flock sites of the shelduck. *Wildfowl*, **32**, 79–87.

Caraco, T. (1980) On foraging time allocation in a stochastic environment. *Ecology*, **61**, 119–28.

Caraco, T. and Wolf, L.L. (1975) Ecological determinants of group sizes for foraging lions. *American Naturalist*, **109**, 343–52.

Carpenter, F.L. and MacMillen, R.E. (1976) Threshold model of feeding territoriality and test with a Hawaiian honeycreeper. *Science*, **194**, 639–42.

Caswell, H. (1989) *Matrix population models*. Sunderland: Sinauer Associates, Inc.

Catterall, C.P., Kikkawa, J. and Gray, C. (1989) Inter-related age-dependent patterns of ecology and behaviour in a population of silvereyes (Aves: Zosteropidae). *Journal of Animal Ecology*, **58**, 557–70.

Cavalli-Sforza, L.L. and Feldman, M.W. (1981) *Cultural transmission and evolution: a quantitative approach*. Princeton: Princeton University Press.

Cayford, J.T. and Goss-Custard, J.D. (1990) Seasonal changes in the size selection of mussels, *Mytilus edulis*, by oystercatchers, *Haematopus ostralegus:* an optimality approach. *Animal Behaviour*, **40**, 609–24.

Ceri, R.D. and Fraser, D.F. (1983) Predation and risk of foraging minnows: balancing conflicting demands. *American Naturalist*, **121**, 552–61.

Charman, K. (1979) Feeding ecology and energetics of the dark-bellied brent goose (*Branta bernicla bernicla*). In R.L. Jefferies and A.J. Davy (ed.), *Ecological processes in coastal environments* (pp. 451–65). Oxford: Blackwells.

Charnov, E.L. (1992) *The theory of sex allocation*. New Jersey: Princeton University Press.

Clark, J.A., Baillie, S.R., Clark, N.A. and Langston, R.H.W. (1993) *Estuary wader capacity following severe weather mortality* BTO Research Report No 103.

Clark, L.R. (1964) Predation by birds in relation to the population density of *Cardiaspina albitextura* (Psyllidae). *Australian Journal of Zoology*, **12**, 349–61.

Clutton-Brock, T.H. (1989) Mammalian mating systems. *Proceedings of Royal Society London Series B*, **236**, 339–72.

Clutton-Brock, T.H., Deutsch, J.C. and Nedft, R.J.C. (1993) The evolution of ungulate leks. *Animal Behaviour*, **46**, 1121–38.

Clutton-Brock, T.H., Green, D., Hiraiwa-Hasegowa, M. and Albon, S.D. (1988) Passing the buck: resource defence, lekking and mate choice in fallow dear. *Behavioural Ecology and Sociobiology*, **23**, 281–96.

Clutton-Brock, T.H., Price, O.F., Albon, S.D. and Jewell, P.A. (1991) Persistent instability and population regulation in soay sheep. *Journal of Animal Ecology*, **60**, 593–608.

Coe, M.J., Bourn, D. and Swingland, I.R. (1979) The biomass, production and carrying capacity of giant tortoises on Aldabra. *Philosophical Transactions of the Royal Society London Series*, **B286**, 163–76.

Cohen, D. (1967) Optimization of seasonal migratory behavior. *American Naturalist*, **101**, 5–17.

Collins, B.T. and Wendt, J.S. (1989) The breeding bird survey in Canada, 1966–1983: analyis of trends in breeding bird populations. *Canadian Wildlife Service Progress Note*, **199**.

Comins, H.N. and Hassell, M.P. (1979) The dynamics of optimally foraging predators and parasitoids. *Journal of Animal Ecology*, **48**, 335–51.

Côté, I.M. and Gross, M.R. (1993) Reduced disease in offspring: a benefit of coloniality in sunfish. *Behavioural Ecology and Sociobiology*, **33**, 269–74.

Courtney, S.P. and Parker, G.A. (1985) Mating behaviour of the tiger blue butterfly (*Tarucus theophtastus*): competive mate-searching when not all females are captured. *Behavioural Ecology and Sociobiology*, **17**, 213–221.

Cox, G.W. (1985) The evolution of avian migration systems between temperate and tropical regions of the new world. *American Naturalist*, **126**, 451–74.

Crawley, M.J. and Krebs, J.R. (1992) Foraging theory. In M.J. Crawley (ed.), *Natural enemies: the population biology of predators, parasites and disease* (pp. 90–114). Oxford: Blackwells.

Curio, E. (1976) *The ethology of predation*. Berlin: Springer-Verlag.

Darwin, C. (1859) *The origin of species*. London: Murray.

Davidson, N. and Rothwell, P. (Eds.). (1993) *Disturbance to waterfowl on estuaries*. Wader Study Group.

Davidson, N.C., Loffoley, D.d., Doody, J.P., Way, L.S., Gordon, J., Key, R. *et al.* (1991) *Nature conservation and estuaries in Great Britain*. Peterborough: Nature Conservancy Council.

Davies, N.B. (1978) Ecological questions about territorial behaviour. In J.R. Krebs and N.B. Davies (ed.), *Behavioural ecology: an evolutionary approach.* (pp. 317–50). Oxford: Blackwell Scientific.

Davies, N.B. (1991) Mating systems. In J.R. Krebs and N.B. Davies (ed.), *Behavioural ecology* (pp. 263–94). Oxford: Blackwell Scientific.

Davies, N.B. (1992) *Dunnock behaviour and social organisation*. Cambridge: Cambridge University Press.

Davies, N.B. (1977). Prey selection and social behaviour in wagtails (Aves: Motacillidae). *Journal of Animal Ecology*, **46**, 37–57.

Davies, N.B. and Green, R.E. (1976) The development and ecological significance of feeding techniques in the reed warbler (*Acrocephalus scirpaceus*). *Animal Behaviour*, **24**, 213–29.

Davies, N.B. and Halliday, T.R. (1979) Competitive mate searching in common toads, *Bufo bufo*. *Animal Behaviour*, **27**, 1253–67.

Davies, N.B. and Houston, A.I. (1983) Time allocation between territories and flocks and owner–satellite conflict in foraging pied wagtails, *Motacilla alba*. *Journal of Animal Ecology*, **52**, 621–34.

Davies, N.B. and Houston, A.I. (1984) Territory economics. In J.R. Krebs and N.B. Davies (ed.), *Behavioural ecology* (2nd edn) (pp. 148–69). Oxford: Blackwell Scientific.

Davis, J.W.F. and O'Donald, P. (1976) Territory size, breeding time and mating preference in the arctic skua. *Nature*, **260**, 774–5.

DeAngelis, D.L., Goldstein, R.A. and O'Neill, R.V. (1975) A model for trophic interaction. *Ecology*, **56**, 881–92.

Deutsch, J.C. (1994) Uganda kob mating success does not increase on larger leks. *Behavioural Ecology and Sociobiology*, **34**, 451–9.

Deutsch, J.C. and Nefdt, R.J.C. (1992) Olfactory cues influence female choice in two lek-breeding antelope. *Nature*, **356**, 596–8.

Deutsch, J.C. and Weeks, P. (1992) Uganda lob prefer high visibility leks and territories. *Behavioural Ecology*, **3**, 223–33.

Dhondt, A.A. (1983) Variations in the number of overwintering stonechats possibly caused by natural selection. *Ringing and Migration*, **4**, 155–8.

Dhondt, A.A. (1988) Carrying capacity: a confusing concept. *Acta Oecologia*, **9**, 337–46.

Dhondt, A.A., Kempenaers, B. and Adriaensen, F. (1992) Density- dependent clutch size caused by habitat heterogeneity. *Journal of Animal Ecology*, **61**, 643–8.

Dill, L.M. and Fraser, A.H.G. (1984) Risk of predation and the feeding behaviour of juvenile coho salmon (*Oncorhynchus kisutch*). *Behavioural Ecology and Sociobiology*, **16**, 65–71.

Dobson, A. and Poole, J.H. (in press) The population dynamics of African elephants and their response to exploitation for ivory. *Conservation Biology*.

Dolman, P.M. and Sutherland, W.J. (1995) The response of bird populations to habitat loss. *Ibis*, **137**, 538–46.

Draulans, D. (1987) The effect of prey density on foraging behaviour and success of adult and first-year grey herons (*Ardea cincerea*). *Journal of Animal Ecology*, **56**, 479–94.

Dunn, E. (1977) Predation by weasels, *Mustela nivalis*, on breeding tits, *Parus* spp., in relation to the density of tits and rodents. *Journal of Animal Ecology*, **46**, 634–52.

Dunn, E.K. (1972) Effect of age on the fishing ability of sandwich terns *Sterna sandvicensis*. *Ibis*, **114**, 360–6.

Durell, L.V., Dit, S.E.A. and Goss-Custard, J.D. (1984) Prey selection within a size class of mussels *Mytilus edulis*, by oystercatchers, *Haematopus ostralegus*. *Animal Behaviour*, **32**, 1197–203.

Eason, P. (1992) Optimization of territory shape in heterogeneous habitats: a field study of the red-capped cardinal (*Paroaria gularis*). *Journal of Animal Ecology*, **61**, 411–24.

East, M. (1988) Crop selection, feeding skills and risks taken by adult and juvenile rooks *Corvus frugilegus*. *Ibis*, **130**, 294–9.

Ebbinge, B. (1991) The impact of hunting on mortality rates and spatial distribution of geese wintering in the Western Palearctic. *Ardea*, **79**, 197–210.

Ehlinger, T.J. (1990) Habitat choice and phenotype-limited feeding efficiencies in bluegill: individual differences and tropic polymorphisms. *Ecology*, **71**, 886–96.

Elgar, M.A. (1986) The establishment of foraging flocks in house sparrows: risk of predation and daily temperature. *Behavioural Ecology and Sociobiology*, **19**, 433–8.

Elliott, J.M. (1989) Mechanisms responsible for population regulation in young migratory trout, *Salmo trutta*. I. The critical time for survival. *Journal of Animal Ecology*, **58**, 987–1001.

Elliott, J.M. (1990) Mechanisms responsible for population regulation in young migratory trout, *Salmo trutta*. III. The role of territorial behaviour. *Journal of Animal Ecology*, **59**, 803–18.

Elliott, J.M. (1994) *Quantitative ecology and the brown trout*. Oxford: Oxford University Press.

Emlen, S.T. (1991) Evolution of cooperative breeding in birds and mammals. In J.R. Krebs, and N.B. Davies (ed.), *Behavioural ecology* (pp. 301–35). Oxford: Blackwell Scientific.

Emlen, S.T. and Oring, L.W. (1977) Ecology, sexual selection and the evolution of mating systems. *Science*, **197**, 215–23.

Ens, B.J. and Goss-Custard, J.D. (1984) Interference among oystercatchers, *Haematopus ostralegus*, feeding on mussels, *Mytilus edulis,* on the Exe estuary. *Journal of Animal Ecology*, **53**, 217–31.

Ens, B.J., Kersten, M., Brenninkmeijer, A. and Hulscher, J.B. (1992) Territory quality, parental effort and reproductive success of oystercatchers (*Haematopus ostralegus*). *Journal of Animal Ecology*, **61**, 703–15.

Ens, B.J., Weissing, F.J. and Drent, R.H. (in press) The despotic distribution and deferred maturity: two sides of the same coin. *American Naturalist*.

Erickson, R.C. and Derrickson, R.R. (1981) The whooping crane. Crane research around the world. In J.C. Lewis and H. Masatomi (ed.), *International crane symposium*, 190 (pp. 104–18). Sapporo.

Espin, P.M.J., Mather, R.M. and Adams, J. (1983) Age and foraging success in black-winged stilts *Himantopus himantopus*. *Ardea*, **71**, 225–8.

Esselink, P. and Zwarts, L. (1989) Seasonal trend in burrow depth and tidal variation in feeding activity of *Nereis diversicolor*. *Marine Ecology Progress Series*, **56**, 243–54.

Essen, L.v. (1991) A note on the lesser white-fronted goose *Anser erythropus* in Sweden and the result of a re-introduction scheme. *Ardea*, **79**, 305–6.

Evans, P.R., Herdson, D.M., Knights, P.J. and Pienkowski, M.W. (1979) Short-term effects of reclamation of part of seal sands, Teesmouth, on wintering waders and shelduck. *Oecologia*, **41**, 183–206.

Ewald, P.W., Hunt, G.L. and Warner, M. (1980) Territory size in western gulls: importance of intrusion pressure, defense investments and vegetation structure. *Ecology*, **61**, 80–7.

Ferguson, S.H., Bergerud, A.T. and Ferguson, R. (1988) Predation risk and habitat selection in the persistence of a remnant caribou population. *Oecologia*, **76**, 236–45.

Fischer, D.H. (1981) Wintering ecology of thrashers in southern Texas. *Condor*, **82**, 392–7.

Fish, P.A. and Savitz, J. (1983) Variations in the home range of largemouth bass, yellow perch, bluegills and pumpkinseeds in an Illinois Lake. *Transactions of the American Fisheries Society*, **112**, 147–53.

Fisher, R.A. (1958) *The genetical theory of natural selection*. London: Dover Publications.

Fitzgerald, G.J., Whoriskey, F.G., Morrissette, J. and Harding, M. (1992) Habitat scale, female cannibalism and male reproductive success in three-spined sticklebacks (*Gasterosteus aculeatus*). *Behavioural Ecology*, **2**, 141–7.

Fitzgibbon, C.D. (1990) Why do hunting chetahs prefer male gazelles? *Animal Behaviour*, **40**, 837–45.

Foster, M.S. (1983) Disruption, dispersion and dominance in lek-breeding birds. *American Naturalist*, **122**, 53–72.

Frank, L.G. (1986) Social organisation of the spotted hyaena (*Crocuta crocuta*) II Dominance and reproduction. *Animal Behaviour*, **35**, 1510–527.

Frank, P.W. (1982) Effects of winter feeding on limpets by black oystercatchers, *Haematopus ostralegus*. *Ecology*, **63**, 1352–62.

Free, C.A., Beddington, J.R. and Lawton, J.H. (1977) On the inadequancy of simple models of mutual interference for parasitism and predation. *Journal of Animal Ecology*, **46**, 543–54.

Freeland, W.J. (1976) Pathogens and the evolution of primate sociality. *Biotropica*, **8**, 12–24.

Freeland, W.J. (1980) Mangabey (*Cercocebus albigena*) movement patterns in relation to food availability and faecal contamination. *Ecology*, **61**, 1297–303.

Fretwell, S.D. (1969) Dominance behavior and winter habitat distribution in juncos (*Junco hyemalis*). *Bird Banding*, **34**, 293–306.

Fretwell, S.D. (1972) *Populations in a seasonal environment*. Princeton: Princeton University Press.

Fretwell, S.D. (1980) Evolution of migration in relation to factors regulating bird numbers. In A. Keast and E.S. Morton (ed.), *Migrant birds in the neotropics: ecology, behavior, distribution and conservation*. Washington: Smithsonian Institution Press.

Fretwell, S.D. (1986) Distribution and abundance of the dickcissel. *Current Ornithology*, **4**, 211–42.

Fretwell, S.D. and Lucas, J.H.J. (1970) On territorial behaviour and other factors influencing habitat distribution in birds. *Acta Biotheoretica*, **19**, 16–36.

Fryxell, J.M. (1991) Forage quality and aggregation by large herbivores. *American Naturalist*, **138**, 478–98.

Fryxell, J.M., Greever, J. and Sinclair, A.R.E. (1988) Why are migratory ungulates so abundant? *American Naturalist*, **131**, 781–98.

Fryxell, J.M. and Sinclair, A.R.E. (1988) Seasonal migration by white-eared kob in relation to resources. *African Journal of Ecology*, **26**, 17–31.

Furness, R.W. (1973) Roost selection by waders. *Scottish Birds*, **7**, 281–7.

Galbraith, H. (1988) The effects of territorial behaviour on lapwing populations. *Ornis Scandinavica*, **19**, 134–8.

Gaston, K.J. and Lawton, J.H. (1987) A test of statistical techniques for detecting density dependence in sequential censuses of animal populations. *Oecologia*, **74**, 401–10.

Gauthreaux, S.A. (1978) The ecological significance of behavioral dominance. In P.P.G. Bateson and P.H. Klopfer (ed.), *Perspectives in ethology* (pp. 17–54). New York: Academic Press.

Gauthreaux, S.A.J. (1982) The ecology and evolution of avian migration systems. In D.S. Farner, J.R. King, and K.C. Parkes (ed.), *Avian biology* (pp. 93–167). New York: Academic Press.

Gibb, J.A. (1958) Predation by tits and squirrels on the eucosmid *Ernarmonia conicolans* (Heyl). *Journal of Animal Ecology*, **27**, 375–96.

Gibb, J.A. (1960) Populations of tits and goldcrests and their food supply in pine plantations. *Ibis*, **102**, 163–208.

Gibson, R.M. (1989) Field playback of male display atracts females in lek breeding sage grouse. *Behavioural Ecology and Sociobiology*, **24**, 439–43.

Gibson, R.M. and Bradbury, J.W. (1986) Male and female strategies on sage grouse leks. In D.I. Rubenstein and R.W. Wrangham (ed.), *Ecological aspects of social evolution* (pp. 379–98). Princeton: Princeton University Press.

Giliam, J.F. and Fraser, D.F. (1987) Habitat selection under predation hazard: test of a model with foraging minnows. *Ecology*, **68**, 1856–62.

Gill, J. A., Sutherland, W. J. and Watkinson, A. R. (in press) A method to quantify the effects of human disturbance on animal populations. *Journal of Applied Ecology*.

Gillis, D.M. and Kramer, D.L. (1987) Ideal interference distributions: population density and patch use by zebrafish. *Animal Behaviour*, **35**, 1875–82.

Gillis, D.M., Kramer, D.L. and Bell, G. (1986) Taylor's power law as a consequence of Fretwell's ideal free distribution. *Journal of Theoretical Biology*, **123**, 281–7.

Giraldeau, L.-A. and Gillis, D. (1985) Optimal group size can be stable. A reply to Sibly. *Animal Behaviour*, **33**, 666–7.

Giraldeau, L.-A., Soos, C. and Beauchamp, G. (1994) A test of the producer–scrounger foraging game in captive flocks of spice finches, *Lonchura punctulata*. *Behaviour Ecology and Sociobiology*, **34**, 251–6.

Glase, J. (1973) Ecology and social organisaiton of the black-capped chickadee. *Living Bird*, **12**, 235–367.

Godin, J.-G.J. and Keenleyside, M.H.A. (1984) Foraging on patchily-distributed prey by a cichlid fish (Teleostei: Cichlidae): a test of the ideal free distribution theory. *Animal Behaviour*, **32**, 120–31.

Gosling, L.M. and Petrie, M. (1990) Lekking in topi: a consequence of satellite behaviour by small males at hotspots. *Animal Behaviour*, **40**, 272–87.

Goss-Custard, J.D. (1969) The winter feeding ecology of the redshank, *Tringa totanus. Ibis*, **111**, 338–56.

Goss-Custard, J.D. (1970) Feeding dispersion in some overwintering wading birds. In J.H.Crook (ed.), *Social behaviour in birds and mammals* (pp. 3–35). London: Academic Press.

Goss-Custard, J.D. (1977a) The ecology of the Wash. III Density related behaviour and the possible effects of a loss of feeding grounds on wading birds (Charadrii). *Journal of Applied Ecology*, **44**, 721–939.

Goss-Custard, J.D. (1977b) Optimal foraging and the size selection of worms by redshank *Tringa totanus* in the field. *Animal Behaviour*, **25**, 10–29.

Goss-Custard, J.D. (1977c) Predator responses and prey mortality in Redshank *Tringa totanus* (L.) and a preferred prey *Corophium volutator* (Pallas). *Journal of Animal Ecology*, **46**, 21–35.

Goss-Custard, J.D. (1980) Competition for food and interference amongst waders. *Ardea*, **68**, 31–52.

Goss-Custard, J.D. (1985) Foraging behaviour of wading birds and the carrying capacity of estuaries. In R.M. Sibly and R.H. Smith (ed.), *Behavioural ecology: ecological consequences of adaptive behaviour* (pp. 169–88). Oxford: Blackwell Scientific.

Goss-Custard, J.D. (1993) The effect of migration and scale on the study of bird populations. *Bird Study*, **40**, 81–96.

Goss-Custard, J.D., Caldow, R.W.G. and Clarke, R.T. (1992) Correlates of the density of foraging oystercatchers *Haematopus ostralegus* at different population sizes. *Journal of Animal Ecology*, **61**, 159–73.

Goss-Custard, J.D., Clarke, R.T. and Durell, S.E.A. (1984) Rates of food intake and aggression of Oystercatchers *Haematopus ostralegus* on the most and least preferred mussel *Mytilus edulis* beds of the Exe Estuary. *Journal of Animal Ecology*, **53**, 233–45.

Goss-Custard, J.D. and Durell, S.E.A. (1990) Bird behaviour and environmental planning: approaches in the study of wader populations. *Ibis*, **132**, 273–89.

Goss-Custard, J.D., Durell, S.E.A., McGrorty, S. and Reading, C.J. (1982*b*) Use of mussel *Mytilus edulis* beds by oystercatchers *Haematopus ostralegus* according to age and population size. *Journal of Animal Ecology*, **51**, 543–54.

Goss-Custard, J.D., Ens, B.J. and Durell, S.E.A. (1982*a*) Individual differences in aggressiveness and food stealing amongst wintering oystercatchers. *Animal Behaviour*, **30**, 917–28.

Goss-Custard, J.D., Kay, D.G. and Blindell, R.M. (1977) The density of migratory and overwintering redshank *Tringa totanus* (L.) and curlew, *Numenius arquata* (L.) in relation to the density of their prey in south-east England. *Estuarine Coastal and Marine Science*, **5**, 497–510.

Goss-Custard, J.D. and Moser, M.E. (1988) Rates of change in the numbers of dunlin wintering in British estuaries in relation to the spread of *Spartina anglica*. *Journal of Applied Ecology*, **25**, 95–109.

Goss-Custard, J.D., West, A.D. and Durell, S.E.A. (1993) The availability and quality of the mussel prey (*Mytilus edulis*) of oystercatchers (*Haematopus ostralegus*). *Netherlands Journal of Sea Research*, **31**, 419–39.

Gosselink, J.G. and Aumann, R.H. (1980) Wetland inventories: wetland losses along the United States coasts. *Zietschrift Geomorphology NF Supplement*, **34**, 173–87.

Götmark, F., Winkler, D.W. and Andersson, M. (1986) Flock-feeding on fish schools increases individual success in gulls. *Nature*, **319**, 589–91.

Grant, J.W.A. and Kramer, D.L. (1990) Territory size as a predictor of the upper limit to population density of juvenile salmonids in streams. *Canadian Journal of Fisheries and Aquatic Sciences*, **47**, 1724–37.

Green, R. (1989) Factors affecting the diet of farmland skylarks *Alauda arvensis*. *Journal of Animal Ecology*, **47**, 913–28.

Green, R.E., Hirons, G.J.M. and Johnson, A.R. (1978) The origin of long-term cohort differences in the distribution of greater flamingos *Phoenicopterus ruber roseus* in winter. *Journal of Animal Ecology*, **58**, 543–55.

Greenberg, R. (1980) Demographic aspects of long-distance migration. In A. Keast and E.S. Morton (eds.), *Migrant birds in the tropics: ecology, behaviour, distribution and conservation* (pp. 493–504). Washington: Smithsonian Institution Press.

Grenfell, B.T., Price, O.F., Albon, S.D. and Clutton-Brock, T.H. (1992) Overcompensation and population-cycles in an ungulate. *Nature*, **355**, 823–6.

Grieg-Smith, P.W. (1985) Winter survival, home ranges and feeding of first year and adult bullfinches. In R.M. Sibly and R.H. Smith (ed.), *Behavioural ecology* (pp. 387–92). Oxford: Blackwell Scientific.

Groves, S. (1978) Age-related differences in Ruddy Turnstone foraging and aggressive behaviour. *Auk*, **95**, 95–103.

Hale, W.G. (1980) *Waders*. London: Collins.

Hamilton, W.D. (1971) Geometry for the selfish herd. *Journal of Theoretical Biology*, **31**, 295–311.

Hammer, O. (1941) Biological and ecological investigations on flies associated with pasturing cattle and their excrement. *Vidensk Meddr dansk naturh Foren*, **105**, 1–257.

Hammerstrom, F. and Hammerstrom, F. (1955) Population density and behaviour in Wisconsin prarie chickens (*Tympanuchus cupida pinnatus*). In *11th International Ornithological Congress*, (pp. 459–66). Bierkhäuser Verlagz Basel.

Hansen, L.P. and Jonsson, B. (1991) Evidence of a genetic component in the seasonal return pattern of Atlantic salmon, *Salmo salar*. *Journal of Fisheries Biology*, **38**, 251–8.

Harper, D.G.C. (1982) Competitive foraging in mallards: 'ideal free' ducks. *Animal Behaviour*, **30**, 575–84.

Hassell, M.P. (1978) *The dynamics of arthropod predator–prey systems*. Princeton: Princeton University Press.

Hassell, M.P. and Anderson, R.M. (1984) Host susceptability as a component in host–parasitoid systems. *Journal of Animal Ecology*, **53**, 611–21.

Hassell, M.P., Lawton, J.H. and May, R.M. (1976) Patterns of dynamical behaviour in single species populations. *Journal of Animal Ecology*, **45**, 471–86.

Hassell, M.P. and May, R.M. (1985) From individual behaviour to population dynamics. In R.M. Sibly and R.H. Smith (ed.), *Behavioural ecology* (pp. 3–32). Oxford: Blackwell Scientific.

Hassell, M.P. and Varley, G.C. (1969) New inductive population model for insect parasites and its bearing on biological control. *Nature*, **223**, 1133–6.

Hayes, F.E. (1987) Intraspecific lekptoparasitism and aggression in young, captive red-eared sliders (*Pseudemys scripta elegans*). *Bulletin of the Maryland Herpitological Society*, **23**, 109–12.

Hegner, R.E. and Emlen, S.T. (1987) Territorial organisation of the white-fronted bee-eater in Kenya. *Ethology*, **76**, 189–222.

Helbig, A.J., Berthold, P., Mohr, G. and Querner, U. (1994) Inheritance of a novel migratory direction in central European blackcaps. *Naturwissenschaften*, **81**, 184–6.

Heyder, R. (1955) Hundert jahre gartenamsel. *Beiträge Vogelkunde*, **4**, 64–81.

Hibino, Y. (1986) Female choice for male gregariousness in a stink bug, *Megacopta punctatissimum*. *Journal of Ethology*, **4**, 91–5.

Hildén, O. (1965) Habitat selection in birds. A review. *Annal Zoologica Fennica*, **2**, 53–70.

Hirons, G. and Thomas, G. (1993) Disturbance on estuaries: RSPB nature reserve experience. *Wader Study Group Bulletin*, **68**, 72–8.

Hobbs, N.T. and Swift, D.M. (1988) Grazing in herds: when are nutritional benefits realised? *American Naturalist*, **131**, 760–4.

Hockin, D., Ounsted, M., Gorman, M., Hill, D., Keller, V. and Barker, M.A. (1992) Examination of the effects of disturbance on birds with reference to its importance in ecological assessments. *Journal of Envioronmental Management*, **36**, 253–86.

Hofer, H. and East, M.L. (1993) The commuting system of Serengeti spotted hyaenas: how a predator copes with migratory prey. I Social organisation. *Animal Behaviour*, **46**, 547–57.

Höglund, J., Montgomerie, R.D. and Widemo, F. (1993) Costs and consequences of the variation in ruff leks. *Behavioural Ecology and Sociobiology*, **26**, 173–80.

Höglund, J. and Robertson, J.G.M. (1990) Female preferences, male decision rules and the evolution of leks in the great snipe *Gallinago media*. *Animal Behaviour*, **40**, 15–22.

Holberton, R.L. (1993) An endogenous basis for differential migration in the dark-eyed junco. *Condor*, **95**, 580–7.

Holbrook, S.J. and Schmitt, R.J. (1988) The combined effects of predator risk and food reward on patch selection. *Ecology*, **69**, 125–34.

Holling, C.S. (1959) Some characteristics of simple types of predation and parasitism. *Canadian Entomologist*, **91**, 385–98.

Holmes, W. (1989) *Grass*. Oxford: Blackwell Scientific.

Holmgren, N. (1995) The ideal free distribution of unequal competitors: predictions from a behaviour-based functional response. *Journal of Animal Ecology*, **64**, 197–212.

Holomuzki, J.R. (1986) Predator avoidance and diel patterns of microhabitat use by larval tiger salamanders. *Ecology*, **67**, 737–48.

Houston, A.I. and McNamara, J.M. (1982) A sequential approach to risk taking. *Animal Behaviour*, **30**, 1260–1.

Houston, A.I. and McNamara, J.M. (1988) The ideal free distribution when competive abilities differ: an approach based on statistical mechanics. *Animal Behaviour*, **36**, 166–74.

Hudson, P.J., Dobson, A.P. and Newbord, D. (1992) Do parasites make prey vulnerable to predation? Red grouse and parasites. *Journal of Animal Ecology*, **61**, 681–92.

Hulscher, J.B. (1973) Burying depth and trematode infection in *Macoma balthica*. *Netherlands Journal of Sea Research*, **6**, 141–56.

Hulscher, J.B. (1982) The oystercatcher as a predator of the bivalve, *Macoma balthica* in the Dutch Wadden Sea. *Ardea*, **70**, 89–152.

Hulscher, J.B. (1989) Mortality and survival of oystercatchers *Haematopus ostralegus* during severe winter conditions. *Limosa*, **62**, 177–81.

Ims, R.A. (1988) Spatial clumping of sexually receptive females induces space sharing among male voles. *Nature*, **335**, 541–3.

Inman, A.J. (1990) Group foraging in starlings—distribution of unequal competitors. *Animal Behaviour*, **40**, 801–10.

Jarman, P.J. (1974) The social organisation of antelope in relation to their ecology. *Behaviour*, **48**, 215–67.

Jarman, P.J. and Wright, S.M. (1993) Macropod studies at Wallaby Crek. IX. Exposure and responses of eastern grey kangaroos to dingoes. *Wildlife Research*, **20**, 833–43.

Johnston, D.W. and Hagan, J.M. (1992) An analysis of long-term breeding bird censuses from eastern deciduous forest. In J.M. Hagan III and D.W. Johnston (ed.), *Ecology and conservation of neotropical migrant landbirds* (pp. 75–84). Washington: Smithsonian Institution Press.

Jones, G. (1987) Selection against large size in the sand martin *Riparia riparia* during a dramatic population crash. *Ibis*, **129**, 274–80.

Jonsson, B. (1982) Diadromous and resident trout *Salmo trutta*: is their difference due to genetics? *Oikos*, **3**, 297–300.

Kaitala, A., Kaitala, V. and Lundberg, P. (1993) A theory of partial migration. *American Naturalist*, **142**, 59–81.

Kawata, M. (1988) Mating success, spatial organisation, and male characteristics in experimental field populations of the red-backed vole *Clethrionomys rufocanus bedfordiae*. *Journal of Animal Ecology*, **57**, 217–35.

Keller, V.E. (1990) The effect of disturbance from roads on the distribution of feeding sites of geese (*Anser brachyrhynchus, A. anser*) wintering in north-east Scotland. *Ardea*, **79**, 229–32.

Kendall, W.A. and Sherwood, R.T. (1975) Palatability of leaves of tall fescue and reed canary grass and some of their alkaloids to meadow voles. *Agronomy Journal*, **67**, 667–71.

Kennedy, M. and Gray, R.D. (1993) Can ecological theory predict the distribution of foraging animals—a critical analysis of experiments on the ideal free distribution. *Oikos*, **68**, 158–66.

Kenward, R.E. (1978) Hawks and doves: factors affecting success and selection in goshawk attacks on wood-pigeons. *Journal of Animal Ecology*, **47**, 449–60.

Kersten, M., Britton, R.H., Dugan, P.J. and Hafner, H. (1991) Flock feeding and food intake in little egrets: the effects of prey distribution and behaviour. *Journal of Animal Ecology*, **60**, 241–52.

Ketterson, E.D. and Nolan, V.J. (1976) Geographic variation and its climatic correlates in the sex ratios of eastern-wintering dark-eyed juncos (*Junco hyemalis hyemalis*). *Ecology*, **57**, 679–93.

Ketterson, E.D. and Nolan, V.J. (1983) The evolution of differential bird migration. *Current Ornithology*, **1**, 357–402.

Kikkawa, J. (1980) Winter survival in relation to dominance classes amongst silvereyes *Zosterops lateralis chlorocephala*, of Heron Island, Great Barrier Reef. *Ibis*, **122**, 437–46.

Kluiver, H.N. (1966) Regulation of a bird population. *Ostrich supplement*, **6**, 389–96.

Knight, F.B. (1958) The effects of woodpeckers on populations of the Englemann spruce beetle. *Journal of Economic Entomology*, **51**, 603–7.

Koene, P. (1978) De Scholekster Aantalseffecten op de Voedselopname. drs thesis, University of Groningen.

Komdeur, J. (1992) Importance of habitat saturation and territory quality for evolution of cooperative breeding in the Seychelles warbler. *Nature*, **358**, 493–5.

Komdeur, J. (1993) Fitness-related dispersal. *Nature*, **366**, 23–4.

Korona, R. (1989) Ideal free distribution of unequal competitors can be determined by the form of competition. *Journal of Theoretical Biology*, **138**, 347–52.

Korona, R. (1990) Travel costs and ideal free distribution of ovipositing female flour beetles, *Tribolium confusum*. *Animal Behaviour*, **40**, 186–7.

Koyama, H. and Kira, T. (1956) Intraspecific competition among higher plants. VIII. Frequency distribuion of individual plant weight as affected by the interaction between plants. *Journal Institute Polytechnic Osaka City University*, **7**, 73–94.

Krebs, J.R. (1970) Regulation of numbers in the great tit, *Parus major*. L. *Journal of Zoological Society of London*, **162**, 317–33.

Krebs, J.R. (1971) Territory and breeding density in the great tit, *Parus major* L. *Ecology*, **52**, 2–22.

Krebs, J.R., Stevens, D.W. and Sutherland, W.J. (1983) Persectives in optimal foraging. In A.H. Brush and G.A. Clark (ed.), *Perspectives in ornithology* (pp. 165–216). New York: Cambridge University Press.

Kruijt, J.P., de Vos, G.J. and Bossema, I. (1972) The arena system of the black grouse. In *International Ornithological Congress*, (pp. 399–423).

Kruuk, H. (1972) *The spotted hyaena*. Chicago: Chicago University Press.

L'Abée-Lund, J.H., Langeland, A., Jonsson, B. and Ugedal, O. (1993) Spatial segregation by age and size in arctic charr: a trade-off between feeding possibility and risk or predation. *Journal of Animal Ecology*, **62**, 160–8.

Lack, D. (1954) *The natural regulation of animal numbers*. Oxford: Clarendon.

Lack, P. (1986) *The atlas of wintering birds in Britain and Ireland*. Calton: T & A D Poyser.

Lank, D.B. and Smith, C.M. (1992) Females prefer larger leks: field experiments with ruff *Philomachus pugnax*. *Behavioural Ecology and Sociobiology*, **30**, 323–9.

Lanyon, S.M. and Thompson, C.F. (1986) Site fidelity and habitat quality as determinants of settlement pattern in male painted buntings. *Condor*, **88**, 206–10.

Larsson, K., Forslund, P., Gustafsson, L. and Ebbinge, B.W. (1988) From the high Arctic to the Baltic: the successful establishment of a barnacle goose *Branta leucopsis* population on Gotland, Sweden. *Ornis Scandinavica*, **19**, 182–9.

Laurie, W.A. and Brown, D. (1990) Population biology of marine iguanas (*Amblyrhynchus cristatus*). III Factors affecting survival. *Journal of Animal Ecology*, **59**, 545–68.

Laursen, K., Gram, I. and Alberto, L.J. (1983) Short-term effect of reclamation on mumbers and distrubution of waterfowl at Hojer, Danish Wadden Sea. In *Proceedings of the Third Nordic Congress of Ornithology*, (pp. 97–118).

Leach, I.H. (1981) Wintering blackcaps in Britain and Ireland. *Bird Study*, **28**, 5–15.

Lemon, W.C. (1991) The fitness consequences of foraging behaviour in the zebra finch. *Nature*, **352**, 153–5.

Lemon, W.C. (1993) Heritability of selectively advantageous foraging behaviour in a small passerine. *Evolutionary Ecology*, **7**, 421–8.

Lessells, C.M. (1985) Parasitoid foraging: should parasitism be density dependent? *Journal of Animal Ecology*, **54**, 27–41.

Lessells, C.M. (1995) Putting resource dynamics into continuous input ideal free distribution models. *Animal Behaviour*, **49**, 487–94.

Lill, A. (1976) Lek behaviour in the golden-headed manakin *Pipra erythrocephala* in Trinidad (West Indies). *Advances in Ethology*, **18**, 1–84.

Lima, S.L. and Dill, L.M. (1990) Behavioural decisions made under the risk of predation: a review and prospectus. *Canadian Journal of Zoology*, **68**, 619–40.

Lindström, E. (1986) Territory inheritance and the evolution of group-living in carnivores. *Animal Behaviour*, **34**, 1825–35.

Łomnicki, A. (1978) Individual differences between animals and the natural regulation of their numbers. *Journal of Animal Ecology*, **47**, 461–75.

Łomnicki, A. (1980) Regulation of population density due to individual differences and patchy environment. *Oikos*, **35**, 185–93.

Łomnicki, A. (1988) *Population ecology of individuals*. Princeton: Princeton University Press.

Lott, D.E. (1991) *Intraspecific variation in the social systems of wild vertebrates*. Cambridge: Cambridge University Press.

Lövei, G.L. (1989) Passerine migration between the Palaearctic and Africa. *Current Ornithology*, **6**, 143–74.

Lowe, V.P.W. (1969) Population dynamics of the red deer (*Cervus elaphus*) on Rhum. *Journal of Animal Ecology*, **38**, 425–57.

Lundberg, P. (1987) Partial bird migration and evolutionarily stable strategies. *Journal of Theoretical Biology*, **125**, 351–60.

Lundberg, P. (1988) On the evolution of partial migration in birds. *Trends in Ecology and Evolution*, **3**, 172–5.

MacArthur, R.H. and Pianka, E.K. (1966) On optimal use of a patchy environment. *American Naturalist*, **100**, 603–9.

McClusky, D.S., Bryant, D.M. and Elliott, M. (1992) The impact of land-claim on macrobenthos, fish and shorebirds on the Forth Estuary, eastern Scotland. *Aquatic Conservation: Marine and Freshwater Ecosystems*, **2**, 211–22.

Macdonald, D.W. (1979) Helpers in fox society. *Nature*, **282**, 69–71.

Mackenzie, A., Reynolds, J.D., Brown, V.J. and Sutherland, W.J. (1995) Variation in male mating success on leks. *American Naturalist*, **145**, 632–51.

McNamara, J.M. and Houston, A.I. (1987) Starvation and predation as factors limiting population size. *Ecology*, **68**, 1515–19.

McNamara, J.M. and Houston, A.I. (1990) State dependent ideal free distributions. *Evolutionary Ecology*, **4**, 298–311.

McNamara, J.M. and Houston, A.I. (1992) Risk-sensitive foraging—a review of the theory. *Mathematical Bulletin*, **54**, 355–78.

McNaughton, S.J. (1984) Grazing lawns: animals in herds, plant form, and coevolution. *American Naturalist*, **124**, 863–86.

Madsen, J. (1985) Impact of disturbance on field utilisation of pink-footed geese in West Jutland, Denmark. *Biological Conservation*, **33**, 53–63.

Madsen, J. (1988) Autumn feeding ecology of herbivorous wildfowl in the Danish Wadden Sea, and impact of food supplies and shooting on movements. *Danish Review of Game Biology*, **13**, 1–32.

Madsen, J. (1993) Experimental wildlife reserves in Denmark: a summary of results. *Wader Study Group Bulletin*, **68**, 23–8.

Magurran, A.E. (1986) Individual differences in fish behaviour. In T. J. Pitcher (ed.), *The behaviour of teleost fishes* (pp. 338–65). Croom Helm.

Major, P. (1978) Predator–prey interactions in two schooling fishes, *Caranx ignobilis* and *Stolephorus purpureus*. *Animal Behaviour*, **26**, 760–77.

Marion, L. (1989) Territorial feeding and colonial breeding are not mutually exclusive: the case of the grey heron (*Ardea cinerea*). *Journal of Animal Ecology*, **58**, 693–710.

May, R.M. (1974) Biological populations with non-overlapping generations: stable points, stable cycles and chaos. *Science*, **186**, 645–7.

Maynard Smith, J.M. (1982) *Evolution and the theory of games*. Cambridge: Cambridge University Press.

Maynard Smith, J.M. and Slatkin, M. (1973) The stability of predator–prey systems. *Ecology*, **54**, 384–91.

Mead, C.T. (1983) *Bird migration*. Feltham: Newnes Books.

Meire, P. and Kuyken, E. (1984) Relations between the distribution of waders and the intertidal benthic fauna of the Oosterschelde, Netherlands. In P.R. Evans, J.D. Goss-Custard, and W.G. Hale (ed.), *Coastal waders and wildfowl in winter* (pp. 57–68). Cambridge: Cambridge University Press.

Meire, P.M. and Ervynck, A. (1986) Are oystercatchers (*Haematopus ostralegus*) selecting the most profitable mussels (*Mytilus edulis*)? *Animal Behaviour*, **34**, 1427–35.

Meire, P.M. and Kuijken, E. (1987) A description of the habitat and wader populations of the Slikken van Vianen (Oosterscelde, The Netherlands) before major environmental changes and some predictions on expected changes. *Le Gerfaut*, **77**, 283–311.

Merkel, I. and Merkel, F.W. (1983) Zum wandertrieb der stare. *Luscinai*, **45**, 63–74.

Metcalfe, J.D., Arnold, G.P. and Webb, P.W. (1990) The energetics of migration by selective tidal stream transport: an analysis for plaice tracked in the North Sea. *Marine Biological Association U.K.*, **70**, 149–62.

Milinski, M. (1979) An evolutionary stable feeding strategy in sticklebacks. *Zeitschrift fur Tierpsychology*, **51**, 36–40.

Milinski, M. (1984) Competitive resource sharing: an experimental test of a learning rule for ESS's. *Animal Behaviour*, **32**, 233–42.

Milinski, M. (1985) Risk of predation taken by parasitised fish under competition for food. *Behaviour*, **93**, 203–16.

Milinski, M. (1986) A review of competitive resource sharing under constraints in sticklebacks. *Journal of Fish Biology*, **29 (Suppl A)**, 1–14.

Milinski, M. (1994) Long-term memory for food patches and implications for ideal free distributions in sticklebacks. *Ecology*, **75**, 1150–6.

Milinski, M. and Parker, G.A. (1991) Competition for resources. In J.R. Krebs and N.B. Davies (ed.), *Behavioural Ecology* (pp. 137–68). Oxford: Blackwell Scientific.

Møller, A.P. (1982) Characteristics of Magpie *Pica pica* territories of varying duration. *Ornis Scandinavica*, **13**, 94–100.

Møller, A.P. (1987) Breeding birds in habitat patches: random distribution of species and individuals? *Journal of Biogeography*, **14**, 225–36.

Møller, A.P. (1988) Nest predation and nest site choice in passerine birds in habitat patches of different size: a study of magpies and blackbirds. *Oikos*, **53**, 215–21.

Møller, A.P. (1989) Population dynamics of a declining swallow *Hirundo rustica* population. *Journal of Animal Ecology*, **58**, 1051–63.

Møller, A.P. (1991) Clutch size, nest predation and distribution of avian unequal competitors in a patchy environment. *Ecology*, **72**, 1336–49.

Monaghan, P. (1980) Dominance and dispersal between feeding sites in the herring gull (*Larus argentatus*). *Animal Behaviour*, **28**, 521–7.

Mönkkönen, M. (1990) Removal of territory holders causes influx of small-sixed intruders in passerine bird communities in northern Finland. *Oikos*, **57**, 281–8.

Morris, D.W. (1989) Density-dependent habitat selection: testing the theory with fitness data. *Evolutionary Ecology*, **3**, 80–94.

Morris, D.W. (1991) Fitness and patch selection by white-footed mice. *American Naturalist*, **138**, 702–16.

Moser, M.E. (1988) Limits to the numbers of grey plovers *Pluvialis squatarola* wintering on British Estuaries: an analysis of long-term population trends. *Journal of Applied Ecology*, **25**, 473–86.

Murton, R.K., Coombs, C.F.B. and Thearle, R.J. (1972) Ecological studies of the feral pigeon *Columba livia* var. II Flock behaviour and social organisation. *Journal of Applied Ecology*, **9**, 875–89.

Murton, R.K., Isaacson, A.J. and Westwood, N.J. (1966) The relationships between wood-pigeons and their clover supply and the mechanism of population control. *Journal of Applied Ecology*, **3**, 55–93.

Murton, R.K., Westwood, N.J. and Isaacson, A.J. (1964) A preliminary investigation of the factors regulating population size in the wood pigeon. *Ibis*, **106**, 482–507.

Myers, J.P. (1981) A test of three hypothesis for lattitudinal segregation of the sexes in wintering birds. *Canadian Journal of Zoology*, **59**, 1527–34.

Myers, J.P., Williams, S.L. and Pitelka, F.A. (1980) An experimental analysis of prey availability for sanderling (Aves: Scolopacidae) feeding on sandy beach crustaceans. *Canadian Journal of Zoology*, **58**, 1564–74.

Nedft, R.J.C. (1992) Lek-breeding in Kafue lechwe. Ph.D. thesis, University of Cambridge.

Neems, R.M., Lazarus, J. and Mclachlan, A.J. (1992) Swarming behaviour in male choronimid midges: a cost–benefit analysis. *Behavioural Ecology*, **3**, 285–90.

Newton, I. (1986) *The Sparrowhawk*. Berkhampsted: T. & A.D. Poyser.

Newton, I. (1992) Experiments on the limitation of bird numbers by territorial behaviour. *Biological Reviews*, **67**, 129–73.

Nicholson, A.J. (1954) An outline of the dynamics of animal populations. *Australian Journal of Zoology*, **2**, 9–65.

Nicholson, A.J. and Bailey, V.A. (1935) The balance of animal populations. Part 1. *Proceedings of the Zoological Society of London*, **3**, 551–98.

Nilsson, S.G. (1987) Limitation and regulation of population density in the nuthatch *Sitta europea* (Aves) breeding in natural cavities. *Journal of Animal Ecology*, **56**, 921–37.

Norton-Griffiths, M. (1968) The feeding behaviour of the oystercatcher. D.Phil. thesis, Oxford University.

O'Connor, R.J. (1986) Dymanic aspects of avian habitat use. In J. Verner, M.L. Morrison, and C.J. Ralph (ed.), *Wildlife 2000: modelling habitat requirements of terrestrial vertebrates* (pp. 235–40). Madison: University of Wisconsin Press.

Oksanen, T., Oksanen, L. and Fretwell, S.D. (1992) Habitat selection and predator–prey dynamics. *Trends in Ecology and Evolution*, **8**, 313.

Orians, G.H. (1969a) Age and hunting success in the brown pelican (*Pelecanus occidentallis*). *Animal Behaviour*, **17**, 316–19.

Orians, G.H. (1969b) On the evolution of mating systems in birds and mammals. *American Naturalist*, **103**, 589–603.

Orians, G.H. and Wittenberger, J.F. (1991) Spatial and temporal scales in habitat selection. *American Naturalist*, **137**, S29–49.

Owen, M. (1971) The selection of feeding site by white-fronted geese in winter. *Journal of Applied Ecology*, **41**, 79–92.

Owen, M. and Black, J.M. (1989) Factors affecting the survival of barnacle geese on migration from the breeding grounds. *Journal of Animal Ecology*, **58**, 603–18.

Park, T. (1962) Beetles, competition and populations. *Science*, **138**, 1369–75.

Parker, G.A. (1970) The reproductive behaviour and the nature of sexual selection in *Scatophaga stercoraria* L. II The fertilisation rate and the spatial and temporal relationships of each sex around the site of mating and oviposition. *Journal of Animal Ecology*, **39**, 205–28.

Parker, G.A. (1974) The reproductive behaviour and the nature of sexual selection in *Scatophaga stercorcaria* L. (Diptera: Scatophagia). IX. Spatial distribution of fertilization rates and evolution of male search strategy within the reproductive area. *Evolution*, **28**, 93–108.

Parker, G.A. (1978) Searching for Mates. In J.R. Krebs and N.B. Davies (eds.), *Behavioural ecology: an evolutionary approach* (pp. 214–44). Blackwells.

Parker, G.A. (1982) Phenotype-limited evolutionarily stable strategies. In King's-College-Sociobiology-Group (ed.), *Current problems in sociobiology*. (pp. 173–201). Cambridge: Cambridge University Press.

Parker, G.A. and Courtney, S.P. (1983) Seasonal incidence: adaptive variation in the timing of life history stages. *Journal of Theoretical Biology*, **105**, 147–55.

Parker, G.A. and Knowlton, N. (1980) The evolution of territory size—some ESS models. *Journal of Theoretical Biology*, **84**, 445–76.

Parker, G.A. and Sutherland, W.J. (1986) Ideal free distribution when individuals differ in competitive ability: phenotype-limited ideal free models. *Animal Behaviour*, **34**, 1222–42.

Peach, W., Baillie, S. and Underhill, L. (1991) Survival of British Sedge Warblers *Acrocephalus schoenobaenus* in relation to west African rainfall. *Ibis*, **133**, 300–5.

Peacock, J.M. (1975) Temperature and leaf growth in four grass species. *Journal of Applied Ecology*, **13**, 225–32.

Percival, S. M., Sutherland, W. J. and Evans, P. R. (in press) A spatial depletion model of the responses of grazing wildfowl to changes in availability of intertidal vegetation during autumn and winter. *Journal of Applied Ecology*.

Perrins, C.M, and Birkhead, T.R. (1983) *Avian ecology*. Glasgow: Blackie.

Peters, W.D. and Grubb, P. (1983) An experimental analysis of sex-specific foraging in the downy woodpecker *Picoides pubescens*. *Ecology*, **64**, 1437–43.

Pianka, E.R. (1978) *Evolutionary ecology*. New York: Harper and Row.

Pienkowski, M.W. and Evans, P.R. (1982) Breeding behaviour, productivity and survival of colonial and non-colonial shelducks *Tanorna tanorna*. *Ornis Scandinavica*, **13**, 101–16.

Piersma, T. and Jukema, J. (1990) Budgeting the flight of a long-distance migrant: changes in nutrient reserve levels of bar-tailed godwits at successive spring staging sites. *Ardea*, **78**, 315–37.

Piersma, T. and van de Sant, S. (1992) Pattern and predictability of potential wind assistance for waders and geese migrating from West Africa and the Wadden Sea to Siberia. *Ornis Svecica*, **2**, 55–66.

Poole, A.F. (1989) *Ospreys: a natural and unnatural history*. Cambridge: Cambridge University Press.

Power, M.E. (1974) Habitat quality and the distribution of algae-grazing catfish in a Panamanian stream. *Journal of Animal Ecology*, **53**, 357–74.

Power, M.E. (1990) Resource enhancement by indirect effects of grazers—armoured catfish, algae and sediment. *Ecology*, **71**, 897–904.

Power, M.E., Mathews, W.J. and Stewart, A.J. (1985) Grazing minnows, piscivorous bass, and stream algae: dynamics of a strong interaction. *Ecology*, **66**, 1448–56.

Pulliam, H.R. and Caraco, T. (1984) Living in groups: is there an optimal group size? In J.R. Krebs and N.B. Davies (ed.), *Behavioural ecology: an evolutionary approach* (pp. 122–47). Oxford: Blackwell Scientific.

Pulliam, H.R. and Danielson, B.J. (1991) Sources, sinks and habitat selection: a landscape perspective on population dynamics. *American Naturalist*, **137**, S50–66.

Ranta, E., Rita, H. and Lindström, K. (1993) Competition versus cooperation: success of individuals foraging alone and in groups. *American Naturalist*, **142**, 42–58.

Rappole, J.H. and Warner, D. (1980) Ecological Aspects of migrant bird behaviour in Veracruz, Mexico. In A. Keast and E.S. Morton (ed.), *Migrant birds in the neotropics: ecology, behavior, distribution and conservation* (pp. 353–95). Washington DC: Smithsonian Institution Press.

Reading, C.J. and McGrorty, S. (1978) Seasonal variation in the burying depth of *Macoma balthica* (L.) and its accessibility to wading birds. *Estuarine, Coastal and Marine Science*, **6**, 135–44.

Recher, H.F. and Recher, J.A. (1969) Comparative foraging efficiency of adult and immature little blue herons *Florida caerula*. *Animal Behaviour*, **17**, 320–2.

Reynolds, J.D., Colwell, M.A. and Cooke, F. (1986) Sexual selection and spring arrival times of red-necked and Wilson's phalaropes. *Behavioural Ecology and Sociobiology*, **18**, 303–10.

Rhijn, L.v. (1983) On maintenance and origin of alternative mating strategies in the ruff *Philomachus pugnax*. *Ibis*, **125**, 482–98.

Robbins, C.S., Fitzpatrick, J.W. and Hamel, P.B. (1992) A warbler in trouble: *Dendroica cerulea*. In J.M. Hagan III and D.W. Johnston (ed.), *Ecology and conservation of neotropical migrant landbirds* (pp. 75–84). Washington: Smithsonian Institution Press.

Robbins, C.S., Sauer, R.S., Greenbeg, S. and Droege, S. (1979) Population declines in North American birds that migrate to the neotropics. *Proceedings of the National Academies of Science*, **86**, 7658–62.

Robinson, S.K., Grzybowski, J.A., Rothstein, S.I., Brittingham, M.C., Petit, L.J. and Thompson, F.R. (1993) Management implications of cowbird parasitism on Neotropical migrant songbirds. In D.M. Finch and P.W. Stangel (ed.), *Status and management of neotropical migrant birds* Rocky Mountain Forest and Range Experimental Station: US. Department of Agiculture.

Rodríguez-Estrella, R. and Rivera-Rodríguez, L. (1992) Kleptoparasitism and other interactions of Crested caracara in the Cape region, Baja California, Mexico. *Journal of Field Ornithology*, **63**, 177–80.

Rohani, P., Godfray, H.C.J. and Hassell, M.P. (1994) Aggregation and the dynamics of host–parasitoid systems: a discrete generation model with within-generation redistribution. *American Naturalist*, **144**, 491–509.

Roland, J., Hannon, S.J. and Smith, M.A. (1986) Foraging pattern of pine siskins and its influence on winter moth survival in an apple orchard. *Oecologia*, **69**, 47–52.

Rose, G.A. and Leggett, W.C. (1990) The importance of scale to predator–prey spatial correlations—an example of atlantic fishes. *Ecology*, **71**, 33–43.

Rosenzweig, M.L. (1981) A theory of habitat selection. *Ecology*, **62**, 327–35.

Rosenzweig, M.L. (1985) Some theoretical aspects of habitat selection. In M.L. Cody (ed.), *Habitat selection in birds* (pp. 517–540). Academic Press, New York.

Rosenzweig, M.L. (1986) Hummingbird isolegs in an experimental system. *Behavioural Ecology and Sociobiology*, **17**, 61–6.

Rosenzweig, M.L. (1991) Habitat selection and population interactions: the search for mechanism. *American Naturalist*, **137**, S5–28.

Rowcliffe, J.M. (1994) The population biology of brent geese and their food plants. Ph.D. thesis, University of East Anglia.

Rowcliffe, J.M., Watkinson, A.R., Sutherland, W.J. and Vickery, J.A. (in press) Cyclical winter grazing patterns in brent geese and the regrowth of *Puccinellia maritima*. *Functional Ecology*.

Royama, T. (1971) Evolutionary significance of predator's response to local differences in prey density. A theoretical study. In P.J. den Boer and G.R. Gradwell (ed.), *Dynamics of populations* (pp. 344–57). Wageningen: Centre for Agricultural Publishing and Documentation.

Rubinstein, D.I. (1981) Individual variation and competition in the everglades pygmy sunfish. *Journal of Animal Ecology*, **50**, 337–50.

Sauer, J.R. and Doege, S. (1990) Recent population trends of the Eastern Bluebird. *Wilson Bulletin*, **102**, 239–52.

Schekkermann, H., Meininger, P.L. and Meire, P.M. (in press) Changes in the waterbird populations of the Oosterschelde, SW. Netherlands, as a result of the large scale coastal engineering works. *Hydrobiologia*.

Schneider, D. (1985) Predation on the urchin *Echinometra lucunter* (Linnaeus) by migratory shorebirds on a tropical reef flat. *Journal of Experimental Marine Biology and Ecology*, **92**, 19–27.

Schneider, D.C. (1978) Equalisation of prey number by migratory shorebirds. *Nature*, **271**, 353–4.

Schneider, D.C. (1992) Thinning and clearing of prey by predators. *American Naturalist*, **139**, 148–60.

Schneider, K.J. (1984) Dominance, predation, and optimal foraging in white-throated sparrow flocks. *Ecology*, **65**, 1820–7.

Schnidrig-Petrig, R., Marbacher, H. and Ingold, P. (1993) Effects of paragliding on the behaviour of chamois. In *18th Ethological Conference*, (pp. 59). Andalucia Gráfica Torremolinos, Spain:

Schoener, T.W. (1983*a*) Field experiments on interspecific competition. *American Naturalist*, **122**, 240–85.

Schoener, T.W. (1983*b*) Simple models of optimal feeding-territory size: a reconciliation. *American Naturalist*, **134**, 608–29.

Schwabl, H. (1983) Ausprägung und bedeutung des Teilzugverhaltens einer süd-westdeutschen Population der Amsel *Turdus merula*. *Journal für Ornithologie*, **124**, 101–16.

Selman, J. and Goss-Custard, J.D. (1988) Interference between foraging redshank *Tringa totanus*. *Ibis*, **36**, 1542–4.

Shelly, T.E. (1990) Waiting for mates: variation in female encounter rates within and between leks of *Drosophila conformis*. *Behaviour*, **107**, 34–48.

Sherman, P.W. (1981) Reproductive competition and infanticide in Belding's ground squirrels and other animals. In R.D. Alexander and D.W. Tinkle (ed.), *Natural selection and social behavior: recent research and new theory* (pp. 311–31). Chiron Press, New York.

Sibly, R.M. (1983) Optimal group size is unstable. *Animal Behaviour*, **31**, 947–8.

Sibly, R.M. and McCleery, R.M. (1983) The distribution between feeding sites of herring gulls breeding at Walney Island. *Journal of Animal Ecology*, **52**, 51–68.

Sih, A. (1984) The behavioural race between predators and prey. *American Naturalist*, **123**, 143–50.

Sinclair, A.R.E. (1975) The resource limitation of tropic levels in tropical grassland ecosystems. *Journal of Animal Ecology*, **44**, 497–520.

Sinclair, A.R.E. (1989) Population regulation in animals. In J. M. Cherrett (ed.), *Ecological concepts* (pp. 197–241). Oxford: Blackwell Scientific.

Slagsvold, T. (1986) Nest site settlement by the pied flycatcher—does the female choose her mate for the quality of his house or himself? *Ornis Scandinavica*, **17**, 210–20.

Smith, H.G. and Nilsson, J.-Å. (1987) Intraspecific variation in migratory pattern of a partial migrant, the blue tit (*Parus caeruleus*): an evaluation of different hypothesis. *Auk*, **104**, 109–15.

Smith, J.N.M. and Arcese, P. (1989) How fit are floaters? Consequences of alternative territorial behaviour in a nonmigratory sparrow. *American Naturalist*, **133**, 830–45.

Smith, P.S. (1975) A study of the winter feeding ecology and behaviour of the bar-tailed godwit (*Limosa lapponica*). Unpublished Ph.D. thesis, University of Durham.

Smith, R.H. and Mead, R. (1974) Age structure and stability in models of predator–prey systems. *Theoretical Population Biology*, **6**, 308–22.

Smith, R.J.F. (1985) *The control of fish migration*. Berlin: Springer-Verlag.

Smith, S.M. (1976) Ecological aspects of dominance hierarchies in black-capped chickadees. *Auk*, **93**, 95–107.

Snow, B.K. and Snow, D.W. (1984) Long-term defence of fruit by mistle thrushes *Turdus viscivorus. Ibis*, **126**, 39–49.

Solomon, M.E. (1949) The natural control of animal populations. *Journal of Animal Ecology*, **18**, 1–35.

Southern, H.N. (1970) The natural control of a population of tawny owls (*Strix aluco*). *Journal of Zoology*, **162**, 197–285.

Stamps, J.A. (1991) The effect of conspecifics on habitat selection in territorial species. *Behavioural Ecology and Sociobiology*, **28**, 29–36.

Stephens, D.W. (1981) The logic of risk-sensitive foraging preferences. *Animal Behaviour*, **29**, 628–9.

Stephens, D.W. and Krebs, J.R. (1986) *Foraging theory*. Princeton: Princeton University Press.

Stillman, R.A., Clutton-Brock, T.H. and Sutherland, W.J. (1992) Black holes, mate retention and the evolution of ungulate leks. *Behavioural Ecology*, **4**, 1–6.

Stock, M. (1993) Studies on the effects of disturbance on staging brent geese: a progress report. *Wader Study Group Bulletin*, **68**, 29–34.

Suhonen, J. (1993) Predation risk influences the use of foraging sites by tits. *Ecology*, **74**, 1197–2103.

Sullivan, K.A. (1989) Predation and starvation: age-specific mortality in juvenile juncos (*Junco phaenotus*). *Journal of Animal Ecology*, **58**, 275–86.

Summers, R.W. (1990) The exploitation of beds of green algae by brent geese. *Estuarine, Coastal and Shelf Science*, **31**, 107–12.

Summers, R.W., Stansfield, J., Perry, S. and Bishop, J. (1993) Utilisation, diet and diet selection by brent geese on saltmarsh. *Journal Zoological Society of London*, **231**, 249–74.

Summers, R.W. and Underhill, L.G. (1991) The growth of the population of dark-bellied brent geese *Branta b. bernicla* between 1955 and 1988. *Journal of Applied Ecology*, **28**, 574–85.

Sutherland, W.J. (1982a) Do oystercatchers select the most profitable cockles? *Animal Behaviour*, **30**, 857–61.

Sutherland, W.J. (1982b) Food supply and dispersal in the determination of wintering population levels of oystercatchers *Haematopus ostralegus. Estuarine, Coastal and Shelf Science.*, **14**, 223–9.

Sutherland, W.J. (1982c) Spatial variation in the predation of cockles by oystercatchers at Traeth Melynog, Anglesey. II. The pattern of mortality. *Journal of Animal Ecology*, **51**, 491–500.

Sutherland, W.J. (1983) Aggregation and the ideal free distribution. *Journal of Animal Ecology*, **52**, 821–8.

Sutherland, W.J. (1992) Game theory models of functional and aggregative responses. *Oecologia*, **90**, 150–2.

Sutherland, W.J. and Allport, G.A. (1994) A spatial model of the interaction between bean geese and wigeon with the consequences for habitat management. *Journal of Animal Ecology*, **63**, 51–9.

Sutherland, W.J. and Anderson, C.W. (1987) Six ways in which a foraging predator may encounter options with different variances. *Biological Journal of the Linnean Society*, **30**, 99–114.

Sutherland, W.J. and Anderson, C.W. (1993) Predicting the distribution of individuals and the consequences of habitat loss: the role of prey depletion. *Journal of Theoretical Biology*, **160**, 223–30.

Sutherland, W.J. and Crockford, N.J. (1993) Factors affecting the feeding distribution of red-breasted geese *Branta ruficollis* wintering in Romania. *Biological Conservation*, **63**, 61–5.

Sutherland, W.J. and Dolman, P.M. (1994) Combining behaviour and population dynamics with applications for predicting the consequences of habitat loss. *Proceedings of the Royal Society London Series B*, **255**, 133–8.

Sutherland, W.J. and Ens, B. (1987) The criteria determining the selection of mussels *Mytilus edulis* by oystercatchers *Haematopus ostralegus*. *Behaviour*, **103**, 187–202.

Sutherland, W.J. and Goss-Custard, J.D. (1991) Predicting the consequence of habitat loss on shorebird populations. In *20th International Ornithological Congress*, (pp. 2199–207).

Sutherland, W.J., Jones, D.W.F. and Hadfield, R.W. (1986) Age differences in the feeding ability of moorhens *Gallinule chloropus*. *Ibis*, **128**, 414–18.

Sutherland, W.J. and Koene, P. (1982) Field estimates of the strength of interference between oystercatchers. *Oecologia*, **55**, 108–9.

Sutherland, W.J. and Parker, G.A. (1985) The distribution of unequal competitors. In R.H. Smith and R.M. Sibly (ed.), *Behavioural ecology: the ecological consequences of adaptive behaviour* (pp. 255–78). Oxford: Blackwell Scientific.

Sutherland, W.J. and Parker, G.A. (1992) The relationship between continuous input and interference models of ideal free distributions with unequal competitors. *Animal Behaviour*, **44**, 345–55.

Sutherland, W.J. and Stillman, R.A. (1988) The foraging tactics of plants. *Oikos*, **52**, 239–44.

Sutherland, W.J., Townsend, C.R. and Patmore, J.M. (1988) A test of the ideal free distribution with unequal competitors. *Behavioural Ecology and Sociobiology*, **23**, 51–3.

Svedarsky, W.D. (1988) Reproductive ecology of female greater prarie chickens in Minnesota. In A.T. Bergerud and M.W. Gratson (ed.), *Adaptive strategies and population ecology of northern grouse. Vol 1 Theory and synthesis* (pp. 193–239). Minneapolis: University of Minnesota Press.

Svensson, B.G. and Pettersson, E. (1994) Mate choice tactics and swarm size: a model and a test in a dance fly. *Behavioural Ecology and Sociobiology*, **35**, 161–8.

Swennen, C. (1984) Differences in quality of roosting flocks of oystercatchers. In P.R. Evans, J.D. Goss-Custard, and W.G. Hale (ed.), *Coastal waders and wildfowl in winter* (pp. 177–89). Cambridge: Cambridge University Press.

Swingland, I.R. (1983) Intraspecific differences in movements. In I.R. Swingland and P.J. Greenwood (ed.), *The ecology of animal movement*. Oxford: Clarendon.

Swingland, I.R. and Lessells, C.M. (1979) The natural regulation of giant tortoise populations on Aldabra Atoll: movement polymorphism, reproductive success and mortality. *Journal of Animal Ecology*, **48**, 639–54.

Székely, T. (1992) Reproduction of kentish plover *Charadrius alexandrinus* in grasslands and fish-ponds: the habitat mal-assessment hypothesis. *Aquila*, **99**, 59–68.

Székely, T. and Bamberger, Z. (1992) Predation of waders (Charadrii) on prey populations: an exclosure experiment. *Journal of Animal Ecology*, **61**, 447–56.

Talbot, A.J. and Kramer, D.L. (1986) Effects of food and oxygen availability on habitat selection by guppies in a laboratory environment. *Canadian Journal of Zoology*, **64**, 88–93.

Tamisier, A. (1974) Etho-ecological studies of teal wintering in the Carmargue (Rhone Delta, France). *Wildfowl*, **25**, 123–33.

Temple, S.A. (1987) Do predators always capture substandard individuals disproportionately from prey populations? *Ecology*, **68**, 669–74.

Terborgh, J. (1990) *Where have all the birds gone?* Princeton: Princeton University Press.

Terrill, S.B. (1990) Food availability, migratory behaviour and population dynamics of terrestrial birds during the non-reproductive season. Avain Foraging: theory, methodology and applications. In M. L. Morrison, C. J. Ralph, J. Verner, & J. R. Jehl (ed.), *Studies in avian biology* (pp. 438–43). Los Angeles: Cooper Ornithological Society.

Terrill, S.B. and Able, K.P. (1988) Bird migration terminology. *Auk*, **105**, 205–6.

Thomas, L. and Martin, H. (in press) The importance of analysis method for breeding bird survey population trend estimates. *Conversation Biology*.

Thompson, D.B.A. (1981) Feeding behaviour of wintering shelduck in the Clyde Estuary. *Wildfowl*, **32**, 88–98.

Thompson, D.B.A. (1984) Foraging economics in flocks of plovers and gulls. Ph.D. thesis, University of Nottingham.

Tocque, K. (1993) The relationship between parasite burdon and host resources in the desert toad (*Scaphiopus couchii*) under natural environmental conditions. *Journal of Animal Ecology*, **62**, 683–93.

Trail, P.W. (1985) Courtship disruption modifies mate choice in a lek breeding bird. *Science*, **227**, 778–80.

Tregenza, T. (1994) Common misconceptions in applying the ideal free distribution. *Animal Behaviour*, **47**, 485–7.

Trivers, R.L. (1972) Parental investment and sexual selection. In B. Cambell (ed.), *Sexual selection and descent of man* (pp. 139–79). Chicago: Aldine.

Tuite, C.H., Hanson, P.R. and Owen, M. (1984) Some ecological factors affecting winter wildfowl distribution on inland waters in England and Wales, and the influence of water-based recreation. *Journal of Applied Ecology*, **21**, 41–62.

Tuljapurkar, S.D. (1982) Population dynamics in variable environments III. Evolutionary dynamics of r-selection. *Theoretical Population Biology*, **21**, 141–65.

van der Have, Nieboer, E. and Boere, G.C. (1984) Age-related distribution of dunlin in the Dutch Wadden Sea. In P.R. Evans, J.D. Goss-Custard, and W.G. Hale (ed.), *Coastal waders and wildfowl in winter* (pp. 160–76). Cambridge: Cambridge University Press.

van Eerden, M.R. (1984) Waterfowl movements in relation to food stocks. In P.R. Evans, J.D. Goss-Custard, and W.G. Hale (ed.), *Coastal waders and wildfowl in winter* (pp. 84–100). Cambridge: Cambridge University Press.

Van Horne, B. (1981) Demography of *Peromyscus maniculatus* populations in serial stages of coastal coniferous forest on southeast Alaska. *Canadian Journal of Zoology*, **59**, 1045–161.

Vehrencamp, S.L. and Bradbury, J.W. (1984) Mating systems and ecology. In J.R. Krebs and N.B. Davies (ed.). *Behavioural ecology* (pp. 251–78). Oxford: Blackwell Scientific

Verner, J. (1964) Evolution of polygamy in the long-billed marsh wren. *Evolution*, **18**, 252–61.

Vickery, J.A. and Sutherland, W.J. (1992) Brent geese: a conflict between conservation and agriculture. *British Crop Protection Council*, **50**, 187–93.

Vickery, J.A., Sutherland, W.J. and Lane, S. (1994) The solutions to the brent goose problem: an economic analysis. *Journal of Applied Ecology*, **31**, 371–82.

Vickery, J.A., Sutherland, W.J., Watkinson, A.R., Lane, S.J. and Rowcliffe, J.M. (in press) Habitat switching by dark-bellied brent geese *Branta b. bernicla* (L) in relation to food depletion. *Oecologia*.

Vøllestad, L.A. and L'Aee-Lund, J.H. (1987) Reproductive biology of stream-spawning roach, *Rutilus rutilus*. *Environmental Biology of Fishes*, **18**, 219–27.

Walker, A.F.G. (1970) The moult migration of Yorkshire Canada Geese. *Wildfowl*, **21**, 99–104.

Wanink, J. and Zwarts, L. (1985) Does an optimally foraging oystercatcher obey the functional response? *Oecologia*, **67**, 98–106.

Watkinson, A.R. and Davy, A.J. (1985) Population biology of salt marsh and sand dune annuals. *Vegetatio*, **62**, 487–97.

Wauters, L. and Dhondt, A.A. (1990) Red squirrel (*Sciurus vulgaris* Linneaus 1758) population dynamics in different habitats. *Zietschrift für Säugetierkunde*, **55**, 161–75.

Wearing, C.H. (1975) Integrated control of apple pests in New Zealand. 2. field estimation of fifth-instar larval and pupal mortalities of codling moth by tagging with cobalt-58. *New Zealand Journal of Zoology*, **2**, 135–49.

Weins, J.A. (1985) Habitat selection in variable environments: shrub-steppe birds. In M.L. Cody (ed.), *Habitat selection in birds* (pp. 227–51). London: Academic Press.

Werner, E.E. (1986) Amphibian metamorphosis: growth rate, predation risk and the optimal time to transform. *American Naturalist*, **128**, 319–41.

Werner, E.E. and Hall, D.J. (1988) Ontogenetic habitat shifts in bluegill: the foraging rate–predation risk trade-off. *Ecology*, **69**, 1352–66.

Werner, E.E., Gilliam, J.F., Hall, D.J. and Mittelbach, G.G. (1983) An experimental test of the effects of predation risk on habitat use in fish. *Ecology*, **64**, 1540–8.

Westoby, M. (1985) Does heavy grazing usually improve the food resource for grazers? *American Naturalist*, **126**, 870–1.

White-Robinson, R. (1982) Inland and saltmarsh feeding by wintering brent geese in Essex. *Wildfowl*, **33**, 113–18.

Whitehead, H. and Hope, P.L. (1991) Sperm whalers off the Galápagos Islands and in the Western North Pacific, 1830–1850: Ideal free whalers? *Ethology and Sociobiology*, **12**, 147–61.

Whitfield, D.P. (1990) Individual feeding specializations of wintering turnstone *Arenaria interpres*. *Journal of Animal Ecology*, **59**, 193–211.

Wiens, J.A. (1989) Spatial scaling in ecology. *Functional Ecology*, **3**, 385–97.

Wilbur, H.M. (1984.) Complex life cycles and community organisation in amphibians. In P.W. Price, C.N. Slobodchikoff, and W.S. Gaud (ed.), *A new ecology: novel approaches to interactive systems* (pp. 196–224). New York: Wiley.

Wilbur, H.M. and Collins, J.P. (1973) Ecological aspects of amphibian metamorphosis. *Science*, **182**, 1305–14.

Wilcove, D.S. (1985) Nest predation in forest tracts and the decline of migratory songbirds. *Ecology*, **66**, 1211–14.

Wiley, R.H. (1991) Lekking in birds and mammals: behavioural and evolutionary issues. *Advances in the Study of Behaviour*, **20**, 201–91.

Williamson, M. (1972) *The analysis of biological populations*. London: Edward Arnold.

Winstanley, D., Spencer, R. and Williamson, K. (1974) Where have all the whitethroats gone? *Bird Study*, **21**, 1–16.

Witter, M.S. and Cuthill, I.C. (1993) The ecological costs of avian fat storage. *Philosophical Transactions of the Royal Society of London Series B*, **340**, 73–92.

Wood, B. (1979) Yellow wagtail *Motacilla flava* migration from West Africa to Europe: pointers towards a conservation strategy for migrants on passage. *Ibis 134 Supplement*, **1**, 66–76.

Wyllie, I. and Newton, I. (1991) Demography of an increasing population of sparrowhawks. *Journal of Animal Ecology*, **60**, 749–66.

Ydenberg, R.C. and Prins, H.H.T. (1981) Spring grazing and the manipulation of food quality by barnacle geese. *Journal of Applied Ecology*, **18**, 443–53.

Zwarts, L. (1976) Density related processes in feeding dispersion and feeding activity of teal (*Anas crecca*). *Ardea*, **64**, 192–209.

Zwarts, L., Blomert, A.-M. and Wanink, J.H. (1992) Annual and seasonal variation in the food supply harvestable by knot *Calidris canutus* staging in the Wadden Sea in late summer. *Marine Ecology Progress Series*, **83**, 129–39.

Zwarts, L. and Drent, R.H. (1981) Prey depletion and regulation of predator density: oystercatchers (*Haematopus ostraegus*) feeding on mussels (*Mytilus edulis*). In N.V. Jones and W.J. Wolf (ed.), *Feeding and survival strategies of esturine organisms* (pp. 193–216). London: Plenum.

Zwarts, L. and Wanink, J. (1984) How oystercatcher and curlews successively deplete clams. In P.R. Evans, J.D. Goss-Custard, and W.G. Hale (ed.), *Coastal waders and wildfowl in winter* (pp. 69–83). Cambridge: Cambridge University Press.

Zwarts, L. and Wanink, J. (1991) The macrobenthos fraction accessible to waders may represent marginal prey. *Oecologia*, **87**, 581–7.

Zwarts, L. and Wanink, J.H. (1993) How the food supply harvestable by waders in the Wadden Sea depends on the variation in energy density, body weight, biomass, burying depth and behaviour of tidal-flat invertebrates. *Netherlands Journal of Sea Research*, **31**, 441–76.

Author index

Subject index